Engineering Optimization: Methods and Applications

Optimization carries great significance in both human affairs and the laws of nature. It refers to a positive and intrinsically human concept of minimization or maximization to achieve the best or most favorable outcome from a given situation. Besides, as the resources are becoming scarce there is a need to develop methods and techniques which will make the systems extract maximum from minimum use of these resources, i.e. maximum utilization of available resources with minimum investment or cost of any kind. The resources could be any, such as land, materials, machines, personnel, skills, time, etc. The disciplines such as mechanical, civil, electrical, chemical, computer engineering as well as the interdisciplinary streams such as automobile, structural, biomedical, industrial, environmental engineering, etc. involve in applying scientific approaches and techniques in designing and developing efficient systems to get the optimum and desired output. The multifaceted processes involved are designing, manufacturing, operations, inspection and testing, forecasting, scheduling, costing, networking, reliability enhancement, etc. There are several deterministic and approximation-based optimization methods that have been developed by the researchers, such as branch-and-bound techniques, simplex methods, approximation and Artificial Intelligence-based methods such as evolutionary methods, Swarm-based methods, physics-based methods, socio-inspired methods, etc. The associated examples are Genetic Algorithms, Differential Evolution, Ant Colony Optimization, Particle Swarm Optimization, Artificial Bee Colony, Grey Wolf Optimizer, Political Optimizer, Cohort Intelligence, League Championship Algorithm, etc. These techniques have certain advantages and limitations and their performance significantly varies when dealing with a certain class of problems including continuous, discrete, and combinatorial domains, hard and soft constrained problems, problems with static and dynamic in nature, optimal control, and different types of linear and nonlinear problems, etc. There are several problem-specific heuristic methods are also existing in the literature.

This series aims to provide a platform for a broad discussion on the development of novel optimization methods, modifications over the existing methods including hybridization of the existing methods as well as applying existing optimization methods for solving a variety of problems from engineering streams. This series publishes authored and edited books, monographs, and textbooks. The series will serve as an authoritative source for a broad audience of individuals involved in research and product development and will be of value to researchers and advanced undergraduate and graduate students in engineering optimization methods and associated applications.

Debashish Das · Ali Safaa Sadiq · Seyedali Mirjalili

Optimization Algorithms in Machine Learning

A Meta-heuristics Perspective

Springer

Debashish Das
Faculty of Computing, Engineering
and the Built Environment
Birmingham City University
Birmingham, UK

Ali Safaa Sadiq
Cyber Security Research Group (CSRG),
Department of Computer Science
Nottingham Trent University
Nottingham, UK

Seyedali Mirjalili
Center for Artificial Intelligence Research
and Optimization
Torrens University Australia
Brisbane, QLD, Australia

University Research and Innovation Center,
Obuda University
Budapest, Hungary

Department of Computer Science
VSB—Technical University of Ostrava
Ostrava, Czech Republic

ISSN 2731-4049 ISSN 2731-4057 (electronic)
Engineering Optimization: Methods and Applications
ISBN 978-981-96-3848-2 ISBN 978-981-96-3849-9 (eBook)
https://doi.org/10.1007/978-981-96-3849-9

This Springer imprint is published by the registered company Springer Nature Singapore Pte Ltd.
The registered company address is: 152 Beach Road, #21-01/04 Gateway East, Singapore 189721, Singapore

Preface

Artificial intelligence which mostly simulates the human intelligence through machine is one of the most trending technologies around the globe that impacts almost every domain these days. Whereas, optimization relates to the process of finding the optimum solution to a particular problem satisfying some given constraints within AI. We decided to write this book to share our understanding of optimization leveraging meta-heuristic algorithms for solving stock market prediction. Owing to its simplicity and flexibility, meta-heuristics have been proven to be effective for solving various optimization problems. To date, there are many meta-heuristics have been developed in the literature. In line with the No Free Lunch theorem which suggests that no single meta-heuristic is the best for all optimization problems, the search for better algorithms is still a worthy endeavor. Grey Wolf Optimizer (GWO) is a meta-heuristic algorithm which is appealing to researcher due to its demonstrated performance as cited in the scientific literature. Despite its merits, GWO is not without limitation. As an example, the current best optimal individual of GWO is biased toward alpha and other individuals (e.g. beta and delta) attempt to modify their positions toward this best individual in each iteration process. This update process may cause the algorithm to fall to local optima especially in the cases where there are many competing local optima. Therefore, the book attempts to explain GWO for improvement of exploration by strengthen the searching process via several random leaders in each iteration, re-generating the random leaders in each iteration and introducing archive to verify the solution with better probability to proceed further for training and re-generation. The verification of each solution individually by modified GWO, instead of considering as a final solution, facilitates the improvement of the exploration. Subsequently, the book attempts to present an ensemble model applying Modified Grey Wolf Optimizer (MGWO) and neural network for stock prediction. Widespread models like Particle Swarm Optimization (PSO), Genetic Algorithm (GA), Ant Colony Optimization (ACO), Evolutionary Strategy (ES), and Population-based incremental learning (PBIL) dealing with the specified problems are also explored and compared in this book. The book presents stock prediction analysis as a case study for training the neural network by adopting MGWO algorithm. In this book, data is used from reputed stock markets; New York Stock Exchange

(NYSE), NASDAQ and emerging markets; Dhaka Stock Exchange (DSE), Bursa Malaysia. Moreover, various factors data like Dollar price, Gold price, Bank interest rate, Foreign Direct Investment, and Inflation are used to measure the effect in stock market. K-means clustering is also explored here in selecting the highly promising company; MGWO can be implemented for feature selection and training; finally, MGWO-NN is explained which can be applied to predict the stock price. The "ensemble" model illustrated here to achieve better predictive performance, can be used to predict future market price. The successful implementation of MGWO and ensemble model is demonstrated in this book. We hope that this book will be useful to you.

This book is an introduction to optimization and application of meta-heuristic algorithm, and it assumes fewer prior knowledge of this field. The first goal of this book is to demonstrate what a meta-heuristic algorithm is and what its applications are. The second goal of this book is to show how to prepare and employ a meta-heuristic algorithm for a given optimization problem: how to create models, how to test them, and how to use them. The final goal of this book is to give an understanding of how to prepare and employ a the meta-heuristic algorithm for solving stock market prediction. In this regard, GWO algorithm is chosen, because it is one of the most well-regarded meta-heuristic algorithms in the literature. This algorithm mimics the leadership hierarchy and hunting mechanism of grey wolves in nature. Besides, the GWO algorithm is applicable to challenging problems in unknown search spaces which produces better result than many other meta-heuristic algorithms. Although the algorithm is free from some limitations, the algorithm can faster decide the suitable thresholds, provide good classification rate, efficiency, and accuracy. This book has 12 chapters that is easy to grasp by any reader.

Birmingham, UK
Nottingham, UK
Brisbane, Australia

Debashish Das
Ali Safaa Sadiq
Seyedali Mirjalili

Acknowledgements

This book would not be possible without the help and support by many people. Having an idea and turning it into a book is as hard as it sounds. The experience is both internally challenging and rewarding. We want to thank God most of all, because without God we wouldn't be able to do any of this. We especially would like to thank every individual that helped make this happen. Complete thanks to Universiti Malaysia Pahang (UMP) and UMP library.

We would like to thank Birmingham City University, UK, for assisting to publish this book. Our warmth thanks also go to friends and colleagues for encouraging, supporting, assisting, and sharing emotions.

Finally, we dedicate this work to our parents. Unforgettable thanks to our families who support and look after and are always beside us.

Contents

About the Authors

Debashish Das is currently teaching at the Faculty of Computer, Engineering and the Built Environment at Birmingham City University, United Kingdom. He obtained his B.Sc., Master's, and Ph.D. degrees in Computer Science in 1999, 2002, and 2019, respectively. With over 22 years of teaching and research experience at prominent universities in the UK, Malta, Malaysia, and Bangladesh, his research interests encompass artificial intelligence, optimization, data science, machine learning algorithms, biomedical applications, and programming languages. He has authored numerous scientific and research articles in reputable national and international journals and conferences.

Ali Safa Sadiq is an associate professor in Cybersecurity and research leader of the Cyber Security Research Group in the Department of Computer Science at Nottingham Trent University. Ali is also a senior IEEE member and adjunct staff at Monash University and an honorary associate professor at the Centre for Artificial Intelligence Research and Optimisation, Torrens University Australia. Ali has served as a senior lecturer in Intelligent Networks at the University of Wolverhampton, and a lecturer at the School of Information Technology, Monash University, Malaysia. Previously he has also served as a senior lecturer at the Department of Computer Systems & Networking Department, Faculty of Computer Systems & Software Engineering, University Malaysia Pahang, Malaysia. Ali completed his first degree in Computer Science in 2004, after that Ali had 5 years of industrial experience in Computer Science and Networking. Ali had his MSc and PhD degrees in Computer Science in 2011 and 2014, respectively. Ali has been awarded the Pro-Chancellor Academic Award as the best student in his batch for both master's and Ph.D. He has published several scientific/research papers in well-known international journals and conferences. He was involved in performing more than 10 research grant projects, where recently with the CSRG team could secured ~£5m research income to perform research projects related to cyber security innovation and development, including the UK national Power Grid supplier OFGEM. Ali has been also involved as a co-investigator with a research project CYBERMIND that was funded £91k by Innovate UK Cyber Academic Start-up Accelerator 2020. Also, he has led (PI) a funded research project called TrustMe, which is funded in two phases by Innovate UK and DCMS. The project creates an innovative new platform to help AI developers and data scientists add security, trust, and explainability to their AI-based decisions. The first phase has been funded with £31,338k, while the second phase was funded with over £60k to develop the proof of concept. Ali managed to develop a commercialized platform called TYMLO and launched a company named TYMLO Technology Ltd. Ali also was part of the project named SOLVD which was funded by the European Regional Development Fund (ERDF) that was supporting local SME companies related to cyber security. He has supervised more than 10 Ph.D. students and 30 Masters students as well as some other undergraduate final year projects. He is currently working on funded projects named Drive with Confidence: A Safe and Secure Driving System to Mitigate Remote Vehicle Hijacking Risks, and PRAVE: PRoactive Authentication and Verification Embedded Model for Critical Cyber-Physical Systems with a total fund of ~£130k. His current research interests include Cybersecurity, Wireless Communications, and AI and optimization algorithms and applications in the Internet of Things and the Internet of Vehicles.

Seyedali Mirjalili is a professor and a globally renowned leader in Artificial Intelligence and Optimization, recognized as the No. 1 AI researcher on Stanford University's prestigious World's Top Scientists list since 2023. He founded the Centre for Artificial Intelligence Research and Optimization in 2019 and serves as a professor of AI at Torrens University Australia, with distinguished professorships in Hungary and the Czech Republic. With over 800 research publications, more than 130,000 citations, and an H-index of 130, Prof. Mirjalili is among the top 1% of highly cited researchers worldwide. His contributions include developing AI algorithms widely applied in science and industry and delivering influential talks, including a TED Talk on AI's transformative potential. Prof. Mirjalili is a strong advocate for responsible and inclusive AI. He has extensively collaborated with industry and government on ethical AI tools. As an editor for leading AI journals, he significantly advances fundamental and applied research in the field. Recognized as a top research leader by The Australian for five years, his insights have earned significant media attention, which showcase his influence as a global thought leader. His groundbreaking work continues to transform industries and drive innovation while addressing societal challenges with ethical and practical solutions.

Abbreviations

ACO	Ant Colony Optimization
ANN	Artificial Neural Network
BA	Bat Algorithm
BBO	Biogeography-Based Optimization
CS	Cuckoo Search
DSE	Dhaka Stock Exchange
EA	Evolutionary Algorithm
ES	Evolutionary Strategies
FA	Firefly Algorithm
GA	Genetic Algorithm
GWO	Grey Wolf Optimizer
HSI	Habitat Suitability Index
MAD	Mean Absolute Deviation
MAPE	Mean Absolute Percentage Error
MGWO	Modified Grey Wolf Optimizer
MLP	Multi-layer Perceptron
NARX	Nonlinear Autoregressive Exogenous
NFL	No Free Lunch
NYSE	New York Stock Exchange
PBIL	Population-Based Incremental Learning
PSO	Particle Swarm Optimization
RMSE	Root Mean Squared Error
RWH	Random Walk Hypothesis
SIV	Suitability Index Variable
SVM	Support Vector Machine

Symbols

A	Shared Archived Score
a_i	Actual Price
$A(k)$	Input of the Neural Network
d	Squared Euclidean Distance
f	Nonlinear Function
$G_\alpha{}^j$	Sum of All the Best Solution
k	Number of Clusters
N	Average Distance between Wolves
p_i	Predicted Price
$P(k)$	Predicted Output
Q_l	Mean
s	Slope
t	Number of Iteration
X	Dataset
$Yn \times k$	Partition Matrix
Π	Vectors

Chapter 1
Challenges and Opportunities in Stock Prediction Using Optimization Techniques

Abstract This chapter introduces the core concept of optimization and its importance in solving scientific, engineering, and industrial problems worldwide. It explores the broad categories of optimization, namely discrete and continuous, and highlights the increasing interest in nature-inspired algorithms for solving complex, real-world problems. Among these, meta-heuristic algorithms, especially the Grey Wolf Optimizer (GWO), have shown promise in areas like classification, learning, and prediction. The chapter discusses the limitations of existing algorithms, outlines the motivation for modifying GWO, and sets the foundation for developing a new ensemble model using a Modified Grey Wolf Optimizer (MGWO) for stock market prediction. It also presents the problem statement, research questions, objectives, scope, and significance of the study, alongside an overview of the book's organization and research framework.

Keywords Optimization · Meta-heuristic algorithms · Grey Wolf Optimizer (GWO) · Ensemble model · Modified Grey Wolf Optimizer (MGWO) · Multi-layer perceptron (MLP)

1.1 Introduction

This chapter exhibits the foundation to the book explaining on the idea of optimization and its pertinence in scientific and industrial procedures everywhere throughout the world. Additionally, the chapter also traces the different sorts of optimization algorithms accessible in literature. Finally, the chapter covers the problem statement, objectives, scope, significance, and the organization of the book.

D. Das et al., *Optimization Algorithms in Machine Learning*, Engineering Optimization: Methods and Applications, https://doi.org/10.1007/978-981-96-3849-9_1

1.2 Optimization in Science and Engineering

Computer science has emerged as a discipline for both theoretical investigation and experimentation, which can solve a seemingly difficult problem perhaps by reducing, embedding, transformation, or simulation. Computer science involves solving problems, designing systems, and understanding human behavior, by drawing on the concepts fundamental to computer science. It often uses massive amounts of data to speed up computation. An exceptionally regular thought in scientific, business, and engineering configuration is the issue of cost and serviceability, in this manner featuring the requirement for optimization. Similarly business associations are worried about expanding benefit, so engineering-design associations are worried about persistently boosting the productivity of the structured items and researchers are consistently looking into to acquire better outcomes with less contribution of time and materials.

There is not really any field of human undertaking today going from medicine, pharmacy, science, engineering to business management that ignores the place of optimization. Optimization is at the core of decision making in manufacturing and mechanical concerns and is a veritable apparatus in the examination of physical frameworks (Gigenrenzer and Gaissmaier 2011). Basically, optimization is dealing about finding the best arrangement out of a few possible arrangements. In scientific terms, optimization manages the look for the ideal object among a few items, particularly in circumstances where an entire feasible search is outlandish (Vazan and Tanuska 2012).

The optimization is such an arena, which can be applicable to attain an optimal solution containing discrete or continuous feasible solutions. Taking all things together, it can be stated that the general goal of either continuous or discrete optimization is to maximize or minimize a function. Alternatively, optimization is the economics of science and engineering with the fact of augmenting benefit, limiting expenses, industrial procedures, or time utilization (Miller and Rubinovich 2012).

Various types of well-known optimizations are available in literature such as Combinatorial Optimization (Wolsey and Nemhauser 2014), Complementarity Problems (Huang and Ni 2010), Constrained Optimization (Bertsekas 2014), Unconstrained Optimization (Tuba et al. 2011), Continuous Optimization (Crandall et al. 2011), Discrete Optimization (Kouvelis and Yu 2013), Global Optimization (Horst and Tuy 2013), Integer Linear Programming (Morais et al. 2010), Linear Programming (LP) (Bazaraa et al. 2011), Network Optimization (Xie et al. 2010), Non-differentiable Optimization Nonlinear Equations (Rao et al. 2012), Optimization Under Uncertainty (Conti et al. 2009), Quadratically Constrained Quadratic Programming (QCQP) (Anstreicher 2012), Quadratic Programming (QP) (Rodriguez-Lujan et al. 2010), Semidefinite Programming (SDP) (Wolkowicz et al. 2012), Semi-infinite Programming (SIP) (Sivaramakrishnan 2002), Stochastic Linear Programming (SLP) (Higle and Sen 2013), Second-Order Cone Programming (SOCP) (Le et al. 2009), Stochastic Programming (Birge and Louveaux 2011), Nonlinear Programming (Kuhn 2014), Nonlinear Least-Squares Problems (Gratton et al. 2007), Mixed Integer Nonlinear Programming (MINLP) (Lee and Leyffer 2011), Bound Constrained Optimization (Morales and Nocedal 2011), Mathematical Programs with Equilibrium Constraints (MPEC) (Luo et al. 1996), Multi-Objective Optimization (Deb 2014), and Derivative-Free Optimization (Rios and Sahinidis 2013). However, this study has categorized these optimizations into two general categories such as discrete and continuous.

Minimizing or maximizing a function using continuous, real numbers by accepting value points from integer set to other set is known as Continuous optimization that contains negative values, decimals, or fractions (Horst and Tuy 2013). So, continuous optimization can take numerical values to make those values appear both in the real world and in the abstract mathematical world. Therefore, some experts believe that continuous optimization is more accurate and complex than its discrete counterpart (Streiner et al. 2014). However, many other experts oppose the finding (Davenport 2013).

Conversely, a subclass of optimization is known as discrete optimization that can use integers as opposed to decimals or fractions and execute minimization or maximization of functions. Combinatorial optimization and integer programming are the two subdivisions of discrete optimization (Nemhauser and Bienstock 2005). Precisely, the current study concentrates on developing nature-inspired optimization algorithm that achieves the solution for continuous or discrete optimization problems stochastically.

In the previous couple of decades in scientific and engineering research, nature-inspired algorithms are becoming progressively prevalent everywhere throughout the world. Researchers are getting excited by this improvement and have illustrated a few purposes behind this: a portion of these causes are that they are created to mimic the best elements in biological, chemical, and physical processes in nature. This circumstance hurls the issue of deciding appropriate algorithm at whatever point a researcher has an optimization issue to solve. Usually, there is a common belief among the researchers that the decision of the 'best' algorithm to tackle a specific issue depends to a great extent on the kind of issue one is faced with. However, there are no such suggested guidelines on a decision of algorithm available for large-scale, nonlinear optimization problems settling (Xu et al. 2012).

Meta-heuristic algorithms are prominent over few decades for solving difficult problems not only in computer science but also in other fields since they are inspired by very simple natural selection concepts. Physical phenomena, animal behaviors, and evolutionary concepts are the typical inspirations of meta-heuristic that facilitates the computer scientist to learn meta-heuristic, simulate various concepts, ensemble meta-heuristic with other algorithms, hybridize one with another, or improve existing meta-heuristic. Hence, the application of meta-heuristic algorithm to solve complex prediction problem consisting nonlinear nature of data is a distinct research area that requires appropriate investigation.

In a nutshell, meta-heuristic algorithms rely on two main components to perform the search process. Exploration is the process of roaming the entire search space to ensure sufficient diversity of the potential solutions. Exploration is the process of exploiting the known best to ensure that the obtained solution is the most optimal.

Excessive exploration tends to increase the computation and may lead to poor convergence. On the other hand, excessive exploitation can make the search process trapped in local optima. For these reasons, there is a need to balance between exploration and exploitation.

Given the aforementioned features, meta-heuristic algorithms can be applied for training neural network even though each algorithm has limitations. Some of the prominent meta-heuristic algorithms include Genetic Algorithm (GA) (Garakani and Branch 2018; Samadzadegan et al. 2010), Particle Swarm Optimization (PSO) (Garakani and Branch 2018; Bao et al. 2013; Blondin and Saad 2010), Bat Algorithm (BA) (Tuba et al. 2016a), Firefly Algorithm (FA) (Tuba et al. 2016b), Cuckoo Search (CS) (Puntura et al. 2016), and Grey Wolf Optimizer (GWO) (Mirjalili et al. 2014b; Eswaramoorthy et al. 2016). However, no heuristic algorithm is the best suited to solve all optimization problems (Yang 2012).

Moreover, limitations of expensive computational cost, occurrence of premature convergence, mutation rate, crossover rate, time-consuming fitness evaluation leads to enhance existing algorithm or propose new one. In machine learning, classification is a supervised learning process to determine appropriate dataset for a new observation based on the performance through training set. Evolutionary or nature-inspired algorithms are good options for classification. Support Vector Machine (SVM) is an efficient supervised learning algorithm that can be applied for classification. The optimization of SVM parameters is possible through algorithms like GA, PSO, BAT, FA, and GWO. The feature selection is a vital part of classification accuracy model and the parameter optimization of SVM through the application of meta-heuristic algorithms, which can simultaneously achieve the feature selection. The feature selection through this process is another extension of distinct research dimension (Wei et al. 2017). However, SVM devises limitations such as computationally expensive, high algorithmic complexity, extensive memory requirements, and selection of appropriate kernel parameters may be tricky (Nika 2015; Patel et al. 2015). Specifically, a problem well handled by a meta-heuristic may not produce same inspiring result for another problem.

The GWO is very efficient for searching that can contribute for classification, feature selection, and learning (Faris et al. 2018; Mirjalili et al. 2014b). For this reason, there is a pressing need to undertake further study to gain complete understanding of the potential offered by this novel algorithm.

This chapter describes the background of study and the challenge statement of the research. This can be within the objectives associated scope of the research to provide an early understanding on the research. The numerous of study and outline of the book organization are going to be outlined in this section. Figure 1.1 shows the overall research mapping components presented in this book, which makes clear idea to the readers of the main aim of this book and the related objectives to it along with the proposed modifications on GWO algorithm that is applied to the stock market data analysis and prediction.

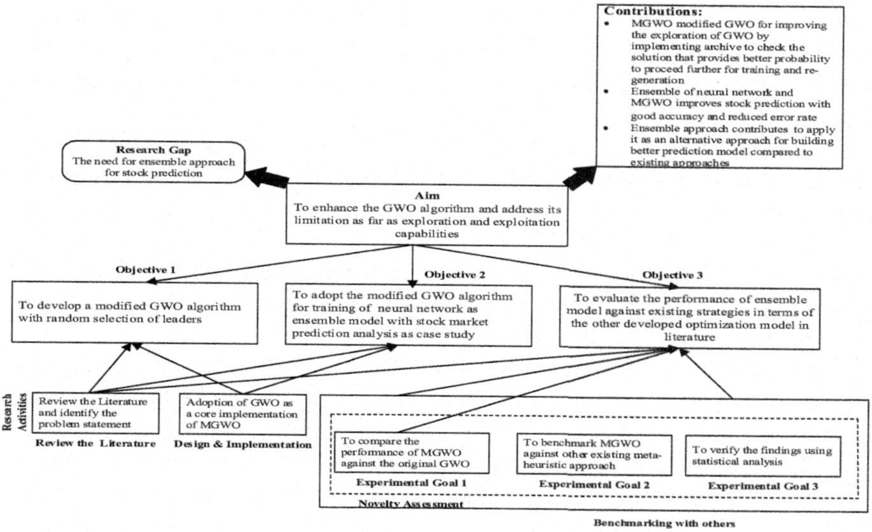

Fig. 1.1 Research mapping

1.3 Problem Statement

In this digital era, huge amount of data is stored and processed all over the world. But, the most challenging task is to extract the useful information from the huge amount of data (Kumar 2014) and hence an appropriate algorithm is required to be developed for exploring the data. Researchers proposed numerous models to achieve good accuracy in prediction through processing large amount of data although no single model is dominant over the other (Nguyen et al. 2015).

Thus, many researchers are fascinated to investigate the area of soft computing due to the higher demand of intelligent system in recent times. A portion of the exceptionally prevalent studies incorporates the Ant Colony Optimization (Dorigo 1992), Bat Algorithm (Yang 2010), Particle Swarm Optimization (Eberhart and Kennedy 1995), and numerous others. These techniques have been effectively implemented to take care of numerous combinatorial issues, for example, Traveling Salesman's Problem, job scheduling, and vehicle routing, just to specify a couple.

Neural network and Support Vector Machine (SVM) are good choices of classifiers for data classification and prediction. However, the accuracy of the prediction depends heavily on the learning that needs proper investigation to determine an appropriate training algorithm (Faris et al. 2018; Wang et al. 2016; Mirjalili et al. 2014a, b). SVM classifier is trained for improvement of classification by applying GWO algorithm by Eswaramoorthy et al. (2016). However, SVM has challenges like high algorithmic complexity, choosing a kernel function is not so easy and long training time for large dataset. Due to the mentioned challenges, this study concentrates on neural network and it training.

Although various heuristic algorithms can be used to train the neural network, the No Free Lunch theorem (NFL) indicates that there is no single meta-heuristic algorithm that is the best suited to solve all optimization problems (Yang 2012) (i.e. to tune the neutral network). For this reason, the investigation of suitable training of neural network is still deemed necessary. One of the very good approaches for classification is through evolutionary or nature-inspired algorithms which originate from the meta-heuristic search algorithms family (i.e. motivated by the theories and biological evolution and the actions of swarms of nature's creation). GWO is one of the recent meta-heuristic algorithms that have demonstrated potential for training neural network and the algorithm can be fine-tuned to perform even better (Faris et al. 2018; Mirjalili et al. 2014a).

The nature provides vast natural wonders with distinct behaviors of animal species. Hence, those unique behaviors and harmonious living of animals can be applied as great inspirations to solve various optimization problems. In this regard, GWO has demonstrated great potential as the algorithm is simple, flexible, derivation-free, and able to avoid local optima. Due to the unique intelligence, GWO algorithm has been modified and applied to solve wide variety of optimization problems compared to other swarm intelligence approach (Faris et al. 2018; Mirjalili et al. 2014b; Nur and Ülker 2018; Turabieh 2016). Some of the successful applications of GWO algorithm to train the neural network include cloud-based intrusion detection and response-based system (Nur and Ülker 2018), prediction of heart disease (Turabieh 2016), melanoma detection (Parsian et al. 2017), design static var compensator controller (Mohamed et al. 2015), and classification of sonar dataset (Mosavi et al. 2016), just to mention a few.

Despite its reported performance, GWO is not without limitations. Specifically, the current best optimal individual is biased toward alpha and other individuals (e.g. beta and delta) attempt to modify their positions toward this best individual in each iteration process. This update process may cause the algorithm to fall into local optima, especially in the cases where there are many competing local optima (Faris et al. 2018; Mirjalili et al. 2014b; Nur and Ülker 2018; Turabieh 2016; Mohamed et al. 2015; Mosavi et al. 2016).

Hence, the proposed research is an attempt to modify GWO to address the deficiency of GWO for improvement of exploration by strengthening the searching process via several random leaders in each iteration, re-generating the random leaders in each iteration, and introducing archive to verify the solution with better probability to proceed further for training and re-generation. The verification of each solution individually by modified GWO, instead of considering as a final solution, facilitates the improvement of the exploration. Moreover, the feature selection restricts number of variables to enhance the performance of the algorithm. With the mentioned approach, the research is an inspiration from the intelligence of modified GWO for prevailing optimization algorithm.

1.4 Research Question

The research questions for this book are:

- Question 1: Can the Grey Wolf Optimizer be enhanced (as modified GWO) to improve its exploration and exploitation capabilities?
- Question 2: Can the ensemble model incorporating MGWO be effectively developed for prediction analysis?
- Question 3: Can the developed ensemble model perform optimally in comparison with existing strategies?

1.5 Aim and Objectives of the Research

The aim of this research is to enhance the GWO algorithm and address its limitation as far as exploration and exploitation capabilities.

The main objectives of the research are:

- To develop a modified GWO algorithm with random selection of leaders.
- To adopt the modified GWO algorithm for training of neural network as ensemble model with stock market prediction analysis as case study.
- To evaluate the performance of ensemble model against existing strategies in terms of the other developed optimization model in literature.

1.6 Scope of the Research

The GWO algorithms models the wolves' navigational ingenuity and implements it to solve optimization problems. The current research work focuses on ensemble intelligent prediction model consisting clustering data mining combined with classification algorithm and neural network that is capable of solving nonlinear problems which can predict stock price trend with significant accuracy using historical stock market prices from the stock market.

The scope of the research is limited to the implementation of MGWO for classification, learning and feature selection. The research will take MGWO algorithm as the core implementation. The research adopts multi-layer perceptron (MLP) neural network trained with MGWO for stock prediction.

1.7 Significance of the Study

The current research will contribute to the common body of knowledge and research in the boundary of Swarm intelligence to take care of optimization issue in industries, engineering, and other genuine issues pertaining to real life. Moreover, the study intends to build up the GWO that will be productive and powerful through persistent exploration and exploitation of the search space. A study of neural network model's efficiency in the selection of models for practical use of stock prediction is another significance of this study.

An ensemble of neural network and MGWO is proposed in this research. GWO is modified for feature selection, classification, and learning by maintaining an archive to select the best solution that provides better probability to proceed further for training and re-generation. This algorithm will be a supervised method where class information needs to be supplied. The algorithms are also be tested with a benchmark dataset.

Neural network performance is enhanced in this research by training it using MGWO to alleviate the problem of over-fitting, entrapment in local minima, result in inaccuracy, and slow convergence rate. In the proposed model, a clustering model is applied to the training dataset followed by a neural network model. This model is regarded as an ensemble model since it combines the neural network and MGWO one after another. A prototype of the ensemble model is implemented to demonstrate its practical use for stock prediction here.

1.8 Research Framework

This section illustrates the complete research activities to attain the research objectives. Precisely, the section illuminates the stages of research development, design, and evaluation of the proposed model as indicated in Fig. 1.2. The research framework is divided into three stages specifically, literature review, research methodology, and evaluation.

1.8.1 Literature Review

Literature Review stage involves reviewing the literature by critically comparing the existing work and analyzing their strengths and limitations in order to justify for the adoption of ensemble model for stock prediction. Moreover, the requirement of the research, theoretical background of GWO, and ensemble model design definitions are established. At this stage, the problem statement is identified and formulated based on the review of existing works.

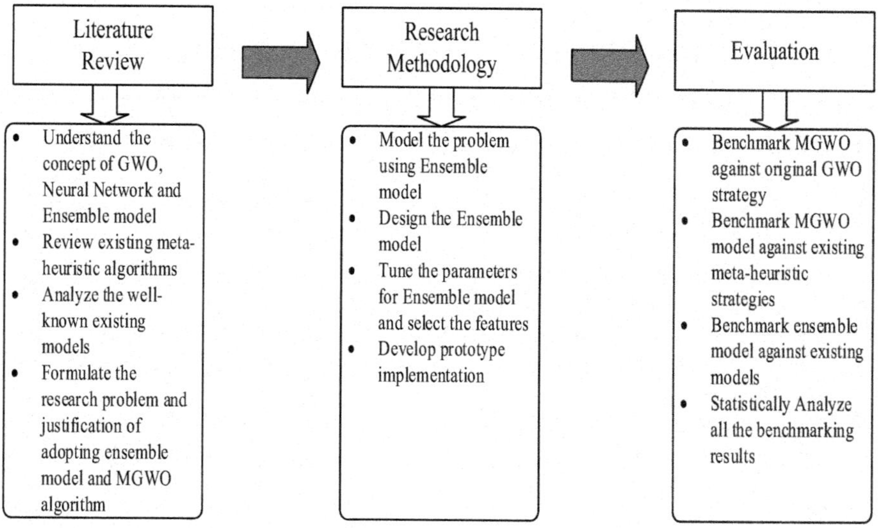

Fig. 1.2 Research framework

1.8.2 Research Activity

Research Activity stage involves finding the best model for stock prediction, and adoption of ensemble model is established. The ensemble model is implemented at this stage applying MGWO and neural network to achieve the best performance. Then, complete algorithm to construct the ensemble model is designed and developed. Additionally, some related concepts such as feature selection, classification, and learning through the application of MGWO are also demonstrated at this stage.

1.8.3 Evaluation

Evaluation stage involves the evaluation of ensemble model. First, MGWO strategy is compared against GWO to evaluate the efficiency of introducing Archive with GWO. Then, MGWO is compared with other meta-heuristic strategies. Next, ensemble model is compared with existing strategies. Finally, the statistical analysis of benchmarking results are performed.

In essence, the book attempts to address three objectives to achieve the aim of the research which is investigation of developing an ensemble model applying MGWO and neural network for stock prediction. Each objective is mapped with several activities that need to be conducted in order to achieve all objectives.

1.9 Book Organization

This book is comprised of eleven chapters. The remaining chapters are organized as:

1. Chapter 2 reviews Optimization methods including deterministic versus stochastic
2. Chapter 3 provides a literature review of heuristic and meta-heuristic optimization algorithms
3. Chapter 4 gives a comprehensive review of Swarm Intelligence methods
4. Chapter 5 presents Artificial Neural Networks
5. Chapter 6 provides a comprehensive review of Grey Wolf Optimizer
6. Chapter 7 includes a survey of Neural Networks trained by the Grey Wolf Optimizer
7. Chapter 8 improves the exploration and exploitation of Grey Wolf Optimizer
8. Chapter 9 shows how to predict stock market using Neural Networks and Grey Wolf Optimizer
9. Chapter 10 extensively compares metaheuristics when training Neural Networks for stock prediction
10. Chapter 11 presents future trends in stock prediction using meta-heuristics.

References

Anstreicher KM (2012) On convex relaxations for quadratically constrained quadratic programming. Math Program 136(2):233–251

Bao Y, Hu Z, Xiong T (2013) A PSO and pattern search based memetic algorithm for SVMs parameters optimization. Neurocomputing 117:98–106

Bazaara MS, Jarvis JJ, Sherali HD (2011) Linear programming and network flows. Wiley

Bertsekas DP (2014) Constrained optimization and Lagrange multiplier methods. Academic Press

Birge JR, Louveaux F (2011) Introduction to stochastic programming. Springer Science & Business Media

Blondin J, Saad A (2010) Metaheuristic techniques for support vector machine model selection. Paper presented at the 10th international conference on hybrid intelligent systems (HIS)

Conti S, Held H, Pach M, Rumpf M, Schultz R (2009) Shape optimization under uncertainty—a stochastic programming perspective. SIAM J Optim 19(4):1610–1632

Crandall D, Owens A, Snavely N, Huttenlocher D (2011) Discrete-continuous optimization for large-scale structure from motion. Paper presented at the 2011 IEEE conference on computer vision and pattern recognition (CVPR)

Davenport TH (2013) Process innovation: reengineering work through information technology. Harvard Business Press

Deb K (2014) Multi-objective optimization search methodologies. Springer, pp 403–449

Dorigo M (1992) Optimization, learning and natural algorithms. Ph.D. thesis, Politecnico di Milano, Italy

Eberhart RC, Kennedy J (1995) A new optimizer using particle swarm theory. Paper presented at the proceedings of the 6th international symposium on micro machine and human science

Eswaramoorthy S, Sivakumaran N, Sekaran S (2016) Grey wolf optimization based parameter selection for support vector machines. COMPEL Int J Comput Math Electr Electron Eng 35(5):1513–1523

Faris H, Aljarah I, Al-Betar MA, Mirjalili S (2018) Grey wolf optimizer: a review of recent variants and applications. Neural Comput Appl 30(2):413–435

Garakani AR, Branch ST (2018) Stock price prediction using multilayer perceptron neural network by monitoring frog leaping algorithm. J Intell Comput 9(1):15–23

Gigenrenzer G, Gaissmaier W (2011) Heuristic decision making. Annu Rev Psychol 62:451–482

Gratton S, Lawless AS, Nicholas NK (2007) Approximate Gauss–Newton methods for nonlinear least squares problems. SIAM J Optim 18(1):106–132

Higle JL, Sen S (2013) Stochastic decomposition: a statistical method for large scale stochastic linear programming, vol 8. Springer Science & Business Media

Horst R, Tuy H (2013) Global optimization: deterministic approaches. Springer Science & Business Media

Huang Z-H, Ni T (2010) Smoothing algorithms for complementarity problems over symmetric cones. Comput Optim Appl 45(3):557–579

Kouvelis P, Yu G (2013) Robust discrete optimization and its applications, vol 14. Springer Science & Business Media

Kuhn HW (2014) Nonlinear programming: a historical view traces and emergence of nonlinear programming. Springer, pp 393–414

Kumar TS (2014) Introduction to data mining. Pearson Education India

Le CV, Gilbert M, Askes H (2009) Limit analysis of plates using the EFG method and second order cone programming. Int J Numer Meth Eng 78(13):1532–1552

Lee J, Leyffer S (2011) Mixed integer nonlinear programming, vol 154. Springer Science & Business Media

Luo Z-Q, Pang J-S, Ralph D (1996) Mathematical programs with equilibrium constraints. Cambridge University Press

Miller BM, Rubinovich EY (2012) Impulsive control in continuous and discrete-continuous systems. Springer Science & Business Media

Mirjalili S, Mirjalili SM, Lewis A (2014a) Let a biogeography-based optimizer train your multi-layer perceptron. Inf Sci 269:188–209

Mirjalili S, Mirjalili SM, Lewis A (2014b) Grey wolf optimizer. Adv Eng Softw 69:46–61

Mohamed AAA, El-Gaafary AA, Mohamed YS, Hemeida AM (2015) Design static VAR compensator controller using artificial neural network optimized by modify grey wolf optimization. Paper presented on 2015 international joint conference on neural networks

Morais H, Kádár P, Faria P, Vale ZA, Khodr HM (2010) Optimal scheduling of a renewable micro-grid in an isolated load area using mixed-integer linear programming. Renew Energy 35(1):151–156

Morales JL, Nocedal J (2011) Remark on "algorithm 778: L-BFGS-B: Fortran subroutines for large-scale bound constrained optimization." ACM Trans Math Softw (TOMS) 38(1):7

Mosavi MR, Khishe M, Ghamgosar A (2016) Classification of sonar data set using neural network trained by gray wolf optimization. Neural Netw World 26(4):393–415

Nemhauser G, Bienstock D (2005) Integer programming and combinatorial optimization. Springer

Nguyen TH, Shirai K, Velcin J (2015) Sentiment analysis on social media for stock movement prediction. Expert Syst Appl 42(24):9603–9611

Nika SS (2015) A comparative study of classification techniques in data mining algorithms. Orient J Comput Sci Technol 8(1):13–19

Nur IM, Ülker E (2018) A hybrid cloud-based intrusion detection and response system (IDRS) based on grey wolf optimizer (GWO) and neural network (NN). Paper presented on 2018 international conference on cyber security and computer science (ICONCS'18)

Parsian A, Ramezani M, Ghadimi N (2017) A hybrid neural network-gray wolf optimization algorithm for melanoma detection. Biomed Res 28(8):3408–3411

Patel J, Shah S, Thakkar P, Kotecha K (2015) Predicting stock and stock price index movement using trend deterministic data preparation and machine learning techniques. Expert Syst Appl 42(1):259–268

Puntura A, Theera-Umpon N, Auephanwiriyakul S (2016) Optimizing support vector machine parameters using cuckoo search algorithm via cross validation. Paper presented at the 6th IEEE international conference on control system, computing and engineering (ICCSCE)
Rao RV, Savsani VJ, Vakharia D (2012) Teaching-learning-based optimization: an optimization method for continuous non-linear large scale problems. Inf Sci 183(1):1–15
Rios LM, Sahinidis NV (2013) Derivative-free optimization: a review of algorithms and comparison of software implementations. J Glob Optim 56(3):1247–1293
Rodriguez-Lujan I, Huerta R, Elkan C, Cruz CS (2010) Quadratic programming feature selection. J Mach Learn Res 11:1491–1516
Samadzadegan F, Soleymani A, Abbaspour RA (2010) Evaluation of genetic algorithms for tuning SVM parameters in multi-class problems. Paper presented at the 11th international symposium on computational intelligence and informatics (CINTI)
Sivaramakrishnan KK (2002) Linear programming approaches to semidefinite programming problems. Renselaer Polytechnic Institute
Streiner DL, Norman GR, Cairney J (2014) Health measurement scales: a practical guide to their development and use. Oxford University Press
Tuba M, Subotic M, Stanarevic N (2011) Modified cuckoo search algorithm for unconstrained optimization problems. Paper presented at the proceedings of the 5th European conference on European computing conference
Tuba E, Tuba M, Simian D (2016a) Adjusted bat algorithm for tuning of support vector machine parameters. Paper presented at the IEEE congress on evolutionary computation (CEC)
Tuba E, Mrkela L, Tuba M (2016b) Support vector machine parameter tuning using firefly algorithm. Paper presented at the 26th international conference on radio-electronics
Turabieh H (2016) A hybrid ANN-GWO algorithm for prediction of heart disease. Am J Oper Res 6(02):136–146
Vazan P, Tanuska P (2012) A short reflection on the strengths and weaknesses of simulation optimization. Paper presented at the proceedings of world academy of science, engineering and technology
Wang GG, Gandomi AH, Zhao X, Chu HCE (2016) Hybridizing harmony search algorithm with cuckoo search for global numerical optimization. Soft Comput 20(1):273–285
Wei Y, Ni N, Liu D, Chen H, Wang M, Li Q, Ye H (2017) An improved grey wolf optimization strategy enhanced SVM and its application in predicting the second major. Math Probl Eng 2017(1):1–12
Wolkowicz H, Saigal R, Vandenberghe L (2012) Handbook of semidefinite programming: theory, algorithms, and applications, vol 27. Springer Science & Business Media
Wolsey LA, Nemhauser GL (2014) Integer and combinatorial optimization. Wiley
Xie C, Lin D-Y, Waller ST (2010) A dynamic evacuation network optimization problem with lane reversal and crossing elimination strategies. Transp Res Part E Logist Transp Rev 46(3):295–316
Xu H, Caramanis C, Mannor S (2012) Sparse algorithms are not stable: a no-free-lunch theorem. IEEE Trans Pattern Anal Mach Intell 34(1):187–193
Yang X-S (2010) A new metaheuristic bat-inspired algorithm. In: Nature inspired cooperative strategies for optimization (NICSO 2010). Springer, pp 65–74
Yang X-S (2012) Swarm-based metaheuristic algorithms and no-free-lunch theorems. In: Theory and new applications of swarm intelligence

Chapter 2
Optimization Methods: Deterministic Versus Stochastic

Abstract Constructing from the preceding chapter, the purpose of this chapter is to provide the critical review of relevant literature to reveal the gaps in literature so that the current research turns into the complement. Initially, the study inspects the concept of optimization and determines the optimization algorithms development including both the stochastic and deterministic algorithms. Moreover, various types of optimization algorithms and their applications, strengths, and weaknesses are analyzed in this chapter. Additionally, the chapter highlights the detail background concepts of the research work, the Grey Wolf Optimizer (GWO), the neural network and its training, and the application of meta-heuristic algorithms for training neural network. The chapter also analyses literature that is relevant to ensemble approach, strength of GWO, limitation, and enhancement of GWO algorithm. Finally, the chapter provides a critical gap analysis in order to justify the proposed research work.

Keywords Optimization · Optimization algorithms · Traditional algorithms · Stochastic algorithms · Nature-Inspired Computing (NIC) · Computing with Nature (CWN)

2.1 Introduction

Chapter 1 includes the brief background and issues related to optimization, neural network and learning, classification. Constructing from the preceding chapter, the purpose of this chapter is to provide the critical review of relevant literature to reveal the gaps in literature so that the current research turns into the complement. Initially, the study inspects the concept of optimization and determines the optimization algorithms development including both the stochastic and deterministic algorithms.

Moreover, various types of optimization algorithms and their applications, strengths, and weaknesses are analyzed in this research. Additionally, the study highlights the detail background concepts of the research work, the Grey Wolf Optimizer

D. Das et al., *Optimization Algorithms in Machine Learning*, Engineering Optimization: Methods and Applications, https://doi.org/10.1007/978-981-96-3849-9_2

(GWO), the neural network and its training, and the application of meta-heuristic algorithms for training neural network. The chapter also analyses literature that is relevant to ensemble approach, strength of GWO, limitation, and enhancement of GWO algorithm. Finally, the chapter provides a critical gap analysis in order to justify the current proposed work.

2.2 Optimization

Industrial and technological advancements are greatly stimulated through optimization all over the world. Optimization strives for better productivity in business, engineering, and manufacturing. Optimization searches for the ideal methods to accomplish an end amidst a few means (Faludi 2013). Fundamentally, optimization includes the maximization or minimization of a function by methodically picking some input values within reasonable set so as to compute the value of the function with the point of deciding the best estimations of the objective function. The general goal of optimization is to guarantee more noteworthy effectiveness utilizing less resources, for example, a computer program could be optimized to utilize less memory, execute quicker, or use minimum resources.

Overall, the objective of optimization technique is to guarantee ideal utilization of accessible resources. However, such facility requires significant cost, for example, a computer program may run quicker and acquire more adequacy, most likely because of its utilization of more memory and the other way around. In general, in this manner, designing of an algorithm is required so that well trade-off between different constraints of an optimization process may be guaranteed. The next section will shed the light on some of the optimization algorithms that represent the state of the arts.

2.3 Optimization Algorithms

The requirement for optimization has stimulated the advancement of correct algorithms, prevalently called deterministic or traditional algorithms, for example, finite volume methods (Said and Wegman 2009), Linear Programming (LP) (Kuhn 2014), Newton–Raphson (Wooldridge 2010), Dynamic Programming (Sniedovich 2010), finite elements (Hughes 2012).

Probabilistic or random elements are not utilized for the proper functioning of Deterministic algorithms (Motwani and Raghavan 2010). Thus, these algorithms yield a similar output values for a given input values and the back-end machines probably could utilize a similar succession of states. In contrast, the stochastic algorithms utilize built-in randomness where, distinctive outcome may be produced by the algorithms for a given set of input values and initial conditions (Gentle 2013; Machairas et al. 2014).

Regardless of this, stochastic models have demonstrated to be very fruitful for comparatively bigger problems consist of numerous input parameters and operating conditions. Alongside, stochastic algorithms have also been implemented recently to establish latest algorithms consist of harmonious and self-organized elements in nature, which is categorized as Natural Computing (Păun 2012).

There are algorithms that basically utilize the computer to generate concepts from nature to create computational frameworks or utilize natural materials, for example, molecules to carry out calculation are identified as Natural Computing. Hence, the definition of Natural Computing includes that nature is the motivation for such computing, at times termed as Nature-Inspired Computing (NIC) or Computing with Natural Materials (CWN) (Dodig-Crnkovic 2012). Natural materials-based computing is the latest achievement in computing approaches where developers make utilization of natural media as a replacement of silicon for computational instruments such as hardware and software (Zang et al. 2010).

Many researchers are enthusiastic about NIC algorithms progressively due to well acceptance in the previous couple of decades in scientific and engineering research everywhere throughout the world. The essential reason given for this prominence is that these algorithms are created to produce the best elements in biological, chemical, and physical procedures in nature (Rozenberg et al. 2011). This circumstance hurls the issue of determining appropriate algorithm as presently a few algorithms are available at whatever point a researcher needs to solve an optimization problem. The decision of a specific algorithm is reliant on its ability to tackle the current problem. This concept is strengthened in optimization by the No Free Lunch theorems (Yang 2011).

Generally, optimization algorithms have the organization as:

$$\text{Minimize} f_i(x)\ (i = 1, 2, 3, \ldots, M), \qquad x \in \Re n \tag{2.1}$$

$$\text{subject to}\ h_a(x) = 0, \qquad (a = 1, 2, 3, \ldots, N), \tag{2.2}$$

$$g_b(x) \leq 0, \qquad (b = 1, 2, 3, \ldots, K) \tag{2.3}$$

where $f_i(x)$, $h_a(x)$, $g_b(x)$ are functions of the design vectors.

$$xiL \leq xi \leq xiU \qquad i = 1, N \tag{2.4}$$

In this occurrence, the function $f_i(x)$ where, $i = 1, 2, \ldots, M$ is known as the objective function. The objective function could be defined as a maximization or minimization problem. For a situation where, $M = 1$, at that point it is an instance of single objective function and for $M \geq 2$, it is a multi-objective function.

Also, the variable $x(i)$ of x is called decision or design variable which could be continuous, discrete, or a blend of both (Feist and Palsson 2010). The space secured by the decision variable is known as the search space $\in \Re$n. Similarly, the space secured by the objective function is known as the solution space, while h_a and g_b are the equality and inequality constraints individually. For the inequality constraints, the maximization has the form ≥ 0, in contrast, the minimization has the form ≤ 0.

In addition, the side constraints are the searchable design space that is characterized by the upper and lower limits, xiL and xiU, of the design or decision variables. Generally, the objective, goal, or cost function can be defined to be linear or nonlinear, implicit or explicit. Integer or discrete optimization problems consist of decision variables with discrete or integer values. On most occasions, conventional optimization technics experience a considerable measure of challenges tackling discrete or integer optimization problems. This is typically the region of solidarity of the stochastic algorithms (Venter 2010).

2.4 Traditional Algorithms

Traditional optimization algorithms are typically deterministic in nature and utilize the gradient-based approach (Davoodi et al. 2014). The example of such algorithms includes the Simplex Method and Newton–Raphson. The traditional optimization algorithms are exceptionally powerful in smooth mono-modal problems because of utilizing functional values and their corresponding derivatives to determine the appropriate result. But, the algorithms may devise the disturbances to the objective function in some situations which deviate the researchers to select non-gradient methods that utilize Hooke-Jeeves pattern search and Nelder–Mead downhill simplex functional values (Haftka and Gürdal 2012).

Substantial number of decision variables is well handled by the traditional optimization algorithms that need limited problem-specific parameter tuning. Moreover, those algorithms are typically ready to get the optimum solution in mono-modal environments. But, traditional optimization algorithms include complex optimization strategies and hence they are not suitable for multimodal search environment.

Additionally, the algorithms experience complexities in taking care of discrete optimization problems and are not so strong in dealing the circumstances like numerical noise (Toga et al. 2012). The stochastic algorithms are required to be developed due to the mentioned shortcomings, which will be discussed further in the following section.

2.5 Stochastic Algorithms

Specifically, two sorts of stochastic algorithms are available such as Nature-Inspired Computing (NIC) and Computing with Nature (CWN). The algorithms can create broad utilization of randomness in searching for optimization (Dodig-Crnkovic 2012). Meta-heuristics also use stochastic components, and because of this stochastic nature, they are especially helpful when the search space is large, multi-modal, or contains discrete decision variables where gradient information is unreliable or unavailable. By alternating global exploration with local exploitation they can escape local minima and discover competitive solutions with relatively modest modelling effort. Nevertheless, they still face open issues such as the absence of rigorous convergence proofs, sensitivity to parameter settings, local optima stagnationn, and high computational demand for very high-dimensional tasks.

2.5.1 Nature-Inspired Computing

Nature-Inspired Computing (NIC) is motivated by the harmonious co-existence and the complex problem-solving techniques of natural environments (Kefi et al. 2015). Consequently, various scientific investigations are motivated by NIC and such investigations are: neural networks (Mäkisara et al. 2014), cellular automata (Codd 2014), artificial immune systems (Hemamalini and Simon 2011), evolutionary computation (Thiele et al. 2009), and swarm intelligence (Ducatelle et al. 2010).

Likewise, robotics researchers are also motivated by nature and proposed mechanical artificial intelligence discipline to develop water strider robot, self-configuring robots, robotic salamander, and mechanical cockroaches (Dewangan et al. 2014). Biologically inspired algorithms are another subset of NIC that can produce the incredible solutions for complex optimization problems through creation of the collective intelligence with a group of biological agents (Pandiri and Singh 2015).

The motivation and development of NIC include the field of biology, chemistry, physics, and engineering. Commonly, NIC systems contain the simulation of harmonious self-organization, interaction, competition, and interdependence of natural elements of the ecosystem. Overall, NIC has been found to acquire answers for issues utilizing heuristics or meta-heuristic standards and this has empowered them to be truly versatile, adaptable, and hearty to the degree that they can be implemented to an extensive variety of utilizations with exceptionally competitive results (Fister et al. 2013).

2.5.2 Computing with Nature (CWN)

Computing with Nature (CWN) transformed computing through the utilization of natural materials replacing silicon. The applications of RNA, DNA, and quantum computing are some of the examples of CWN-based computational processing. Moreover, CWN-based computing are also applied in recent times to molecular or biocomputing, biochemical computing, biomolecular computing or DNA computing that utilizes components from molecular biology for data processing operations such as, logical, arithmetic, and other computer operations (Rozenberg et al. 2011). CWN-based molecular computing has also been implemented effectively to take care of a 7-vertice TSP issue by only exploring different avenues regarding DNA strands in a test tube, 20-variable 3SAT issues, cryptography, sticker frameworks, joining frameworks, and the structure applications for savvy drugs (de Castro 2007).

Alternatively, quantum computing executes computations through the consideration of data as quantum bits and involving mechanical means, for example, entrapments and superpositioning. A quantum bit, which is also referred as qubit, contains a '0', '1' or a quantum superposition of either a '0' or '1'. Additionally, the quantum computer utilizes logic gates to carry out computing operations on the qubits with the guide of Shor's polynomial algorithm for integers factoring and Grover's algorithm for quantum database query (Hirvensalo 2013). Although quantum computing is still in its earlier stages of advancement, researchers are enthusiastic to investigate the true ability of this computing paradigm as quantum computing has demonstrated its potential to quantum cryptography, quantum teleportation, nuclear magnetic resonance imaging, pattern recognition, and classification (Hirvensalo 2013).

2.6 Summary

In this chapter, a quick overview was presented on the nature of optimization algorithms and their two main types of deterministic versus stochastic behavior. The main concept of Linear Programming and Dynamic Programming was discussed to introduce the readers on the initial process of solving such optimization problem. Besides, very generic minimization optimization problem formulation was given to familiarize the readers on the problem formulation process.

On the other hand, some of the nature-inspired stochastic algorithms were discussed and their main categories as were identified in the state of the art, biology, chemistry, physics, and engineering-based NIC. Finally, the concept of Computing with Nature (CWN) was discussed as an introduction to the main topic of this book that deals with grey wolves behavior and its use for constructing an optimization algorithm. The next chapter will review and discuss in more details the heuristic and meta-heuristic optimization algorithms.

References

Codd EF (2014) Cellular automata. Academic Press

Davoodi E, Hagh MT, Zadeh SG (2014) A hybrid improved quantum-behaved particle swarm optimization-simplex method (IQPSOS) to solve power system load flow problems. Appl Soft Comput 21:171–179

de Castro LN (2007) Fundamentals of natural computing: an overview. Phys Life Rev 4(1):1–36

Dewangan S, Naik A, Agrawal A (2014) Study of nature inspired computing

Dodig-Crnkovic G (2012) Info-computationalism and morphological computing of informational structure. In: Integral biomathics. Springer, pp 97–104

Ducatelle F, Di Caro GA, Gambardella LM (2010) Principles and applications of swarm intelligence for adaptive routing in telecommunications networks. Swarm Intell 4(3):173–198

Faludi A (2013) A reader in planning theory. Elsevier
Feist AM, Palsson BO (2010) The biomass objective function. Curr Opin Microbiol 13(3):344–349
Fister Jr I, Yang X-S, Fister I, Brest J, Fister D (2013) A brief review of nature-inspired algorithms for optimization. arXiv preprint arXiv:1307.4186
Gentle JE (2013) Random number generation and Monte Carlo methods. Springer Science & Business Media
Haftka RT, Gürdal Z (2012) Elements of structural optimization, vol 11. Springer Science & Business Media
Hemamalini S, Simon SP (2011) Dynamic economic dispatch using artificial immune system for units with valve-point effect. Int J Electr Power Energy Syst 33(4):868–874
Hirvensalo M (2013) Quantum computing. Springer
Hughes TJ (2012) The finite element method: linear static and dynamic finite element analysis. Courier Corporation
Kefi S, Rokbani N, Krömer P, Alimi AM (2015) A new ant supervised-PSO variant applied to travelling salesman problem. Paper presented at the hybrid intelligent systems: 15th international conference HIS 2015 on hybrid intelligent systems, Seoul, South Korea, 16–18 Nov 2015
Kuhn HW (2014) Nonlinear programming: a historical view traces and emergence of nonlinear programming. Springer, pp 393–414
Machairas V, Tsangrassoulis A, Axarli K (2014) Algorithms for optimization of building design: a review. Renew Sustain Energy Rev 31:101–112
Mäkisara K, Simula O, Kangas J, Kohonen T (2014) Artificial neural networks, vol 2. Elsevier
Motwani R, Raghavan P (2010) Randomized algorithms. Chapman & Hall/CRC
Pandiri V, Singh A (2015) Swarm intelligence approaches for multidepot salesmen problems with load balancing. Appl Intell 1–13
Păun G (2012) Membrane computing. In: Handbook of natural computing, pp 1355–1377
Rozenberg G, Bck T, Kok JN (2011) Handbook of natural computing. Springer Publishing Company, Incorporated
Said Y, Wegman E (2009) Roadmap for optimization. Wiley Interdiscip Rev Comput Stat 1(1):3–17
Sniedovich M (2010) Dynamic programming foundations and principles. CRC Press
Thiele L, Miettinen K, Korhonen PJ, Molina J (2009) A preference-based evolutionary algorithm for multi-objective optimization. Evol Comput 17(3):411–436
Toga AW, Clark KA, Thompson PM, Shattuch DW, Van Horn JD (2012) Mapping the human connectome. Neurosurgery 71(1):1
Venter G (2010) Review of optimization techniques. In: Encyclopedia of aerospace engineering
Wooldridge JM (2010) Econometric analysis of cross section and panel data. MIT Press
Yang X-S (2011) Review of meta-heuristics and generalised evolutionary walk algorithm. Int J Bio-Inspired Comput 3(2):77–84
Zang H, Zhang S, Hapeshi K (2010) A review of nature-inspired algorithms. J Bionic Eng 7(4):S232–S237

Chapter 3
Heuristic and Meta-heuristic Optimization Algorithms

Abstract This chapter introduces and explains heuristic and meta-heuristic optimization algorithms in detail. It highlights their characteristics, differences, and classifications, helping readers understand how these algorithms work to find optimal solutions in complex problem spaces. The chapter describes important features such as exploration, exploitation, and randomization, and explains how different algorithms achieve a balance between these to improve performance. Both trajectory-based and population-based meta-heuristics are discussed, including Simulated Annealing, Hill Climbing, and the Great Deluge algorithm, along with their strengths, limitations, and pseudocode. The chapter also explores the popularity of population-based approaches like Genetic Algorithms, Particle Swarm Optimization, and Grey Wolf Optimizer due to their effectiveness in exploring large search spaces using multiple agents. Overall, the chapter lays a solid foundation for understanding how meta-heuristics can solve real-world optimization problems across various domains.

Keywords Heuristic and meta-heuristic algorithms · Randomization · Exploration · Exploitation · Trajectory-based algorithms · Simulated Annealing · Hill Climbing algorithm · The Great Deluge · Population-based meta-heuristic algorithms

3.1 Introduction

In this chapter, the heuristic and meta-heuristic optimization algorithms will be presented and discussed in more detailed form. The characteristics of these algorithms will be also highlighted along with illustrated pseudocode of each category of these algorithms to give the reader a comprehensive understanding on their process in optimizing the searching process for the optimal solutions in the searching space of the give problem.

D. Das et al., *Optimization Algorithms in Machine Learning*, Engineering Optimization: Methods and Applications, https://doi.org/10.1007/978-981-96-3849-9_3

3.2 Heuristic and Meta-heuristic Algorithms

NIC utilizes heuristic and meta-heuristics algorithms extensively to enrich computation. A complex problem can be solved by heuristic algorithm through the exploitation of some information. Although heuristic algorithms are near-exact algorithms that may not produce exact optimal solution, the utilization of heuristic algorithms yet can produce excellent outcome for complex optimization problems at an appropriate time (Safari 2015). Alternatively, meta-heuristics algorithms, which are also termed as 'beyond heuristics', can act superior to heuristics by utilizing intelligent memory; experiential and different biases to assist manage the search process (Prakasam and Savarimuthu 2015).

Generally speaking, meta-heuristic algorithms apply local search besides global explorations utilizing randomizations, which assist these algorithms with steering far from being limited in a local environment to a progressively global search. The general goal of any meta-heuristic algorithm is to accomplish the most ideal outcome by utilizing typical mechanisms to accomplish satisfactory exploration and exploitation of the search space (Blum and Roli 2003). Extensively, meta-heuristics algorithms can be applied to wide range of areas such as bioinformatics, telecommunications, economics, manufacturing, and so on (Osman and Kelly 2012).

Meta-heuristic algorithms can be commonly categorized into two types, namely population-based and trajectory-based (Behesti and Shamsuddin 2013). Population-based meta-heuristic algorithms can be recognizable to Holland who published his work in 1962 and whose works utilized a mix of theoretical genetics and automata approach. Researchers were inspired to meta-heuristic algorithms because of applying variety and diversification strategies to a population to accomplish results inside a search space. A portion of these approaches can be mentioned as: Schaffer's Vector-Evaluated Genetic Algorithm (VEGA) (Pierre et al. 2011); Farmer, Packard, and Pearson's Artificial Immune Systems (Farmer et al. 1986); Holland's and Rosenberg's Evolutionary Strategies (Cuomo et al. 2012); Dorigo and Di Caro's Ant Colony Optimization ACO (Di Caro and Dorigo 1998); and Grey Wolf Optimizer (Mirjalili et al. 2014a).

3.3 Characteristics of Meta-heuristic Algorithms

A decent meta-heuristic algorithm contains two vital features such as capabilities to engage global search mechanism or exploration and local search or exploitation (Osaba et al. 2016). Where, 'Local Search' can explore the capable neighboring regions in the hope to determine the optimal solution that is termed as exploitation and 'Global Search' facilitates to skip any local optimum that is also referred as exploration.

The efficiency of meta-heuristic algorithm may be substantially adjusted by balancing the interaction between local search or exploitation and global search or exploration. However, searching locally a lot may lead the algorithm to be trapped in local optimum, on the other hand, an aggressive global searching may result in inefficiency that affects the whole performance of the search (Yang et al. 2014).

Another key component of a decent algorithm is the capacity to recognize the best outcome in iteration and conceivably the best structure vector related with such best outcome. Generally, the rule is identified as 'The survival of the fittest'. The criterion can be accomplished by continuously refreshing the present best found up until this point (Yang 2011).

3.4 Randomization in Meta-heuristic

Meta-heuristic algorithms contain three key components, namely exploration, exploitation, and determining the top performer where, each algorithm is differentiated from one another based on the mechanism engaged to attain the mentioned key components (Li et al. 2010). The incorporation of randomization and a deterministic procedure facilitates the meta-heuristic algorithms to attain the goal, where randomization is a mechanism to define the upper and lower boundaries in a uniformly distributed variable ranging from 0 to 1. Particle Swarm Optimization (PSO) and Firefly Algorithm (FA) have adopted this approach.

Cuckoo Search is another strategy used by meta-heuristic that adopts Lévy flight, which is a random process, categorized by step-jumps to look out the disorganized dust particles movement in a fluid (Senthilnath et al. 2012). However, exploration is attained by engaging crossover and mutation for some algorithms such as Genetic Algorithm (GA), Genetic Programming (GP), and Evolutionary Programming (EP). Where, mutation determines the latest solution from initial population and crossover emphasizes the limit on over-exploration (Rani et al. 2012).

The exploitation is achieved for the meta-heuristic algorithms by producing different solutions from initial solution. But, the exploitation is attained for Simulated Annealing (SA) algorithm through engaging random walk (Kirkpatrick et al. 1983) and for Harmony Search (HS) algorithm by pitch alteration (Mahdavi et al. 2007). Equation (3.1) denotes the mentioned approach:

$$X\text{new} = X\text{old} + sw \tag{3.1}$$

where Xnew is the new solution, Xold is the initial solution, s is the step size, and w is zero mean determined from Gaussian distribution. However, the step size should not be too narrow or too wide because too wide step size will support exploration eliminating exploitation and too narrow step size will support exploitation to produce the result trapped into local minima. Therefore, the algorithms should engage Lévy flight or random walk so that the appropriate step size can be determined from Lévy distribution (Kennedy 2010).

3.5 Broad Classification of Meta-heuristic Algorithms

In literature, meta-heuristic algorithms are classified in numerous ways and one such way is population-based or trajectory-based. Determining the solution of a problem by implementing the population of solutions at a period is known as population-based algorithm (Wong and Moin 2015). Population-based algorithm determines the initial solution randomly and then iteratively improves the solution. The algorithms are also known as exploration-based approaches due to the algorithms excellent ability of the search space diversification.

GA and PSO are the two perfect examples of the mentioned approach where, GA utilizes a set of strings and PSO utilizes a number of particles (Kennedy 2010). Alternatively, trajectory-based algorithm utilizes a single agent to revolve around the search space in zigzag manner iteratively and the example of such approaches includes Simulated Annealing, Great Deluge, and Hill Climbing (Kennedy 2010; Mirjalili et al. 2014b). Population-based and trajectory-based algorithms are different from each other based on number of temporal solutions during each search iteration course, where population-based algorithms utilize multiple agents to produce multiple solutions but trajectory-based algorithms utilize a single agent to produce single solution.

3.6 Trajectory-Based Algorithms

Trajectory-based algorithms which are also referred as exploitation-based algorithms initially determine a single solution for the current search and then the solution is improved iteratively to produce the final solution (Park et al. 2013). The algorithms generally emphasize the intensification where, the optimal solution is determined by the search agent that moves through the search space to trace the path in the search landscape (Manjarres et al. 2013). Figure 3.1 demonstrates the pseudocode for trajectory-based algorithms, where the search agents move randomly from one solution to another continuously till the stopping criteria are met in a solution space.

The productivity of trajectory-based algorithm with regard to time and quality of solutions can be enhanced by incorporating parallelism, where parallel multi-start, parallel evaluation, and parallel moves are three well-known parallel models (Alba et al. 2005). In literature, various trajectory-based algorithms are available such as Simulated Annealing, Hill Climbing, and Great Deluge.

```
Begin
Generate initial solution (s(0))
t = 0
While (not Termination (s(t)))
        Explore neighborhood s'(t) = SelectMove(s(t))
        If Move(s'(t)) accepted
            s(t) = ApplyMove (s'(t))
            t = t + 1
        End If
End While
Output best result
End
```

Fig. 3.1 Trajectory-based algorithms pseudocode

3.7 Simulated Annealing

Kirkpatrick et al. established the idea of Simulated Annealing (SA) algorithm to model the cooling and heating processes of materials in metallurgical engineering. The idea of the SA algorithm comprises that the metals become too strong by gradually reducing the temperature so that the system energy can be minimized through the cooling process. The algorithm initiates a random search at high temperature for the cooling process to produce greedy decent till the temperature turns to zero. The algorithm performs well in lower temperature compared to higher because the randomization feature facilitated by the SA algorithm ensures avoiding local optima because the greedy descent may place the algorithm stack into local minima (Dowsland and Thompson 2012).

During each move, SA algorithms search for optimal solutions through the implementation of random variables so that the improvement of the objective function can be determined, where lower objective value is preferable for a minimization problem. Hence, SA algorithms improve the objective function through the avoidance of being trapped to local minima so that global exploration can be maintained. As the algorithms progress, the annealing schedule consisting of both linear and geometric features ensures the effectiveness by reducing the temperature.

Moreover, SA algorithm maintains the search area minimization and earlier convergence features through the reduction of the temperature. Figure 3.2 indicates the pseudocode of SA algorithm that can be implemented for problem minimization (Li and Wei 2008).

SA algorithm can be implemented to solve various problems, for example, Artificial Neural Networks Training (Ledesma et al. 2008), Quadratic Assignment Problem (Bilbao and Alba 2009), Job Shop Scheduling (Van Laarhoven et al. 1992), Traveling Salesman's Problems (Malek et al. 1989), N-Queens Problem (Tambouratzis 1997). SA algorithm is having the capability to avoid local minima through the implicit manipulation of the temperature cooling which is the main strength of the algorithm.

However, SA algorithm may not be very efficient to implement for smooth energy landscape and also for the problems with few local minima. Additionally, SA algorithm may not reach for appropriate outcome within certain period of time because

```
Begin
Initialize population and parameters
Generate randomly an initial optimal state Si,
Calculate f(Si)
Select an initial temperature T0
Select a terminal temperature Tf or a total number of temperature chant tmax
Set temperature change counter t = 1
While Ti < Tf or t = tmax
        Set repetition counter L = 0
        Repeat Until L = Lt = βt
        Generate new state Sj, a neighbor of Si
        Calculate ΔE = f(Sj) - f(Si)
        If ΔE < 0 then
              Si = Sj
        Else If random (0,1) < Exp(-ΔE/(KbT)), then
              Si = Sj where Kb is Bolzmann's constant
        End If
              L = L + 1
        End Repeat
              t = t + 1
              Ti = αTi, where α is the cooling rate
  End While
  End
```

Fig. 3.2 SA algorithms pseudocode

of including many cost function evaluations iteratively (Kumbharana and Pandey 2013).

3.8 Hill Climbing Algorithm

Hill Climbing (HC) algorithms iteratively determine the solution for a problem by picking an arbitrary solution initially and modify the single solution to determine a better solution (Hoffmann 2010). The modification will continue to determine a new solution until no further improvement is possible. But an extra modification on HC algorithm's effort will be attempted to determine the optimal solution if, any modification leads poor solution.

There are numerous ways by which HC algorithm is different from similar algorithms like Gradient Descent, such as HC algorithms fine-tune only a single value for the current solution but Gradient Descent modifies multiple values for current solution for the subsequent iteration. Hence, HC algorithms are considered as a type of Depth-First search (Fisher 1987).

However, HC algorithms apply feedback mechanism to estimate the closeness or latest solution so that the next search direction can be determined, which is different from Depth-First search that rejects or accepts the solution out-rightly. The pseudocode for HC algorithm is demonstrated in Fig. 3.3 that can be applied for the minimization problem. HC algorithms can be implemented in various arenas to obtain

```
Begin
    Initialize population and parameters
    x_k = LB_k + (LB_k - LB_k) x U_(0,1) , k = 1,2, ... N ( The initial state of x)
    Calculate f(x)
    While (until termination condition)
        x' = improved (x)
        If f(x') ≤ f(x) then
            x = x'
        End If
    End While
End
```

Fig. 3.3 Hill Climbing algorithms pseudocode

effective outcomes such as configuring application servers (Xi et al. 2004), Traveling Salesman Problem (Selman and Gomes 2006), and signature verification (Galbally et al. 2007).

HC algorithm is useful compared to other similar algorithms due to lower consumption of computer resources for searching as the algorithms store only current solution. In addition, the algorithms produce better results in comparison with other algorithms for the unexpected interruption during the execution of the algorithm. But, HC algorithms may have some limitations such as low speed for ridges instance, alleys, and plateau, probability of getting stuck into local minima (Minton et al. 1992). However, the issue of slower speed and getting stuck into local minima is the main concern for many trajectory algorithms and hence many researchers successfully investigated to minimize the issue (Sharma et al. 2016).

3.9 The Great Deluge

The Great Deluge (GD) algorithm is proposed by Dueck and includes the concept of a person's activities to move in various directions upwards to the hill during a deluge so that his feet can be avoided to become wet if the water level rises (Özcan et al. 2012). GD algorithms initially assign a value similar to the initial objective function for the parameter and the value is decreased iteratively during the progression of the search. The algorithms produce the final solution if the determined value is nearly equivalent to the objective function.

The GD algorithms have been advanced later by allowing the algorithms to receive all downhill moves and also hybridizing GD with Hill Climbing for better effectiveness (Burke and Bykov 2017). The GD algorithms can be applied by choosing an optimum solution from an approximate solution J. Later, the algorithms select K as a random value of *badness* so that the desired approximate solution can be estimated. This way, J will produce adverse solution for greater assessment of the *badness* value. The algorithms implement another parameter called *tolerance* that can assess numerous factors to select J' as an approximate solution for a neighbor J. The calculation for J' solution's *badness* is determined at this phase to compare the outcome with *tolerance* parameter.

GD algorithms initiate recursively for any outcome better than *tolerance*. But, any outcome worse than *tolerance* will result to choose J'' as a neighboring solution for J that will allow the process to be continued till better results than *tolerance* are determined for all neighbors of J. Finally, GD algorithm will be concluded with a final solution J (Dhouib 2010). Figure 3.4 demonstrates the pseudocode for GD algorithm that can be applied for minimizing a problem (Nabeel 2010; Othman et al. 2013).

GD algorithms can be applied to various extents such as prediction for protein structure (Burke et al. 2007), problems of facility layout (Nahas et al. 2010), issues of patient admission (Kifah and Abdullah 2015), course timetabling, sports, examination, and other similar areas. The algorithms are unlike Hill Climbing and Simulated Annealing due to receive neighborhood candidate solution. GD algorithms are more effective compared to HC and SA as the algorithms allow to explain two characteristics earlier for a search process such as processing time and processing region of the estimated solution (Burke et al. 2003). However, the algorithms are having limitation

```
Begin
Initialize population and parameters
Generate random solution Sol
Set SolutG = Sol
Set Level = f(SolutG)
Set estimated quality of Solution: Estimated.Quality
Set number of iteration: IterNo
Calculate increase rate β = Estimated.Quality / IterNo
Set Iteration = 0
While (iteration < IterNo)
        Generate a random new Solution Sol_new in the neighborhood of Sol
        Calculate f(Sol_new)
        If f(Sol_new) > f(SolutG) then
              Sol_new = SolutG
              Sol = Sol_new
              f(Sol) = f(Sol_new)
              f(SolG) = f(Sol_new)
        End If
        If f(Sol_new) >Level then
              Sol = Sol_new
              f(Sol) = f(Sol_new)
        End If
        If f(Sol_new) ≤ Level then
              Level = Level + β
        End If
        Iteration = Iteration + 1
End While
Return SolutOptimalGD
End
```

Fig. 3.4 Great Deluge algorithms pseudocode

such as possibility of being trapped to local minima that create the variations of the algorithm (Mcmullan 2007). In spite of GD algorithms advancement with variations, the efficiency of the algorithms is yet a main concern.

Overall, various limitations of trajectory algorithms such as possibility of getting trapped to local minima, slower speed due to accept single solution at a time, and inefficient for mixed-objective optimization to maximize or minimize objectives (Yang et al. 2006) motivate the researchers to opt for population-based approaches which accept multiple-optimal solutions at one iteration due to the application of multiple search agents.

3.10 Population-Based Meta-heuristic Algorithms

Population-based approaches usually implement a set of decision vectors, which can be expressed by Eq. (3.2), where N denotes the size of population and the number of design variables is denoted by n (Kothari 2012).

$$X = \{x_1, x_2, x_3, \ldots x_N\} \tag{3.2}$$

where,

$$Xi = (x_i1, x_i2, \ldots, x_in) \tag{3.3}$$

Meta-heuristic term was first introduced and applied by Glover (1986), who has also proposed an algorithm called TS (1989). Escaping from local optima is the main endeavor of Meta-heuristic algorithm that can explore the search space proficiently by trial-and-error basis. The majority of meta-heuristic algorithms are considered as population-based algorithms, where finding a result begins from numerous places of solution space which is different from traditional algorithms.

Consequently, every individual of the population may be the candidate to determine the optimal solution. Meta-heuristic algorithms can guide the searching and movement in the search space which can facilitate to proficiently explore the complete search space and the same may not be possible by other search algorithms. The guidance for search is problem specific, which can be referred as fitness function, where parameters can be maximized or minimized through the fitness function in light of issue nearby.

Population-based meta-heuristics algorithms generally share a similar structure consisting of four components such as main algorithm, extension to deal with constrained optimization problems, extension to retain promising solutions, and a component to halt the program. However, the key algorithms implement three features, namely crossover, mutation, and selection for most cases (Nozohour-Leilabady and Fazelabdolabadi 2016). Figure 3.5 represents the pseudocode for population-based meta-heuristic algorithms (Karaboga and Bastruk 2007).

```
Begin
    Initialize population and parameters
    Evaluate the objective function
    While (until termination condition)
        Evaluate the population quality
        Apply the variation operator
        Evaluate the objective function
    End While
    Output best result
End
```

Fig. 3.5 Population-based algorithms pseudocode

In the literature, many useful meta-heuristic algorithms have been proposed over the last few decades. Some of the most popular meta-heuristic algorithms implemented, but not limited to, are: PSO (Cura 2009; Seidy 2016; Li et al. 2014), GA (Delnavaz 2014; Razali and Geraghty 2011), ACO (Mohapatra and Das 2013; Yang et al. 2014), ES (Wen et al. 2015; Bliss et al. 2014; Bisoi and Dash 2014), PBIL (Ali et al. 2014), BBO (Mirjalili et al. 2014a), and GWO (Mirjalili et al. 2014b).

'Local Search' and 'Global Search' are the two key components of meta-heuristic algorithm. Where, 'Local Search' can explore the capable neighboring regions in the hope to determine the optimal solution that is termed as exploitation and 'Global Search' facilitates to skip any local optimum that is also referred as exploration. The efficiency of meta-heuristic algorithm may be substantially adjusted by balancing the interaction between local search or exploitation and global search or exploration. However, searching locally a lot may lead the algorithm to be trapped in local optimum, on the other hand, an aggressive global searching may result in inefficiency that affects the whole performance of the search (Yang et al. 2014).

Population-based meta-heuristic algorithms usually exploit the prior knowledge from the solution space and the search agent is moved toward the feasible region by utilizing the solution at the initialization phase. If the necessary information is unavailable, the distribution of the decision vectors is taken place uniformly at the search space (Wong and Moin 2015). The algorithms are generally inspired by harmonious co-existence of nature such as bioinspired and swarm-based; however, some algorithms like Grey Wolf Optimizer, Biogeography-based Optimization, Black Hole Optimization, and Harmony Search are inspired by physics, chemistry, or geography.

3.11 Summary

This chapter reviewed several heuristics and meta-heuristics. The main algorithm covered are Simulates Annealing, Hill Climbing, and Great Deluge. There were also discussions around local and global search methods.

References

Alba E, Talbi E, Luque G, Melab N (2005) Parallel metaheuristics: a new class of algorithms. Wiley, pp 79–104

Ali BAS, Ahmed ABED, Muhammad AED (2014) Incremental learning approach for enhancing the performance of multi-layer perceptron for determining the stock trend. Int J Sci Basic Appl Res (IJSBAR) 16(1):15–23

Behesti Z, Shamsuddin SMH (2013) A review of population-based meta-heuristic algorithms. Int J Adv Soft Comput Appl 5(1):1–35

Bilbao M, Alba E (2009) Simulated annealing for optimization of wind farm annual profit. Paper presented at the logistics and industrial informatics

Bisoi R, Dash PK (2014) A hybrid evolutionary dynamic neural network for stock market trend analysis and prediction using unscented Kalman filter. Appl Soft Comput 19:41–56

Bliss CA, Frank MR, Danforth CM, Dodds PS (2014) An evolutionary algorithm approach to link prediction in dynamic social networks. J Comput Sci 5(5):750–764

Blum C, Roli A (2003) Metaheuristics in combinatorial optimization: overview and conceptual comparison. ACM Comput Surv (CSUR) 35(3):268–308

Burke EK, Bykov Y (2017) The late acceptance hill-climbing heuristic. Eur J Oper Res 258(1):70–78

Burke E, Bykov Y, Newall J, Petrovic S (2003) A time-predefined approach to course timetabling. Yugosl J Oper Res 13(2):19–51

Burke E, Bykov Y, Hirst J (2007) Great deluge algorithm for protein structure prediction

Cuomo CA, Desjardins CA, Bakowski MA, Goldberg J, Ma AT, Becnel JJ, Levin JZ (2012) Microsporidian genome analysis reveals evolutionary strategies for obligate intracellular growth. Genome Res 22(12):2478–2488

Cura T (2009) Particle swarm optimization approach to portfolio optimization. Nonlinear Anal Real World Appl 10(4):2396–2406

Delnavaz B (2014) Forecasting of the stock price index by using fuzzy-neural network and genetic algorithms. J Appl Sci Agric 9:109–117

Dhouib S (2010) A multi start great deluge metaheuristic for engineering design problems. Paper presented at the 2010 IEEE/ACS international conference on computer systems and applications (AICCSA)

Di Caro G, Dorigo M (1998) AntNet: distributed stigmergetic control for communications networks. J Artif Intell Res 317–365

Dowsland KA, Thompson JM (2012) Simulated annealing handbook of natural computing. Springer, pp 1623–1655

Farmer JD, Packard NH, Perelson AS (1986) The immune system, adaptation, and machine learning. Phys D 22(1):187–204

Fisher DH (1987) Knowledge acquisition via incremental conceptual clustering. Mach Learn 2(2):139–172

Galbally J, Fierrez J, Ortega-Garcia J (2007) Bayesian hill-climbing attack and its application to signature verification. In: Advances in biometrics. Springer, pp 386–395

Glover F (1986) Future paths for integer programming and links to artificial intelligence. Computers & operations research 13(5):533–549

Hoffmann J (2010) A heuristic for domain independent planning and its use in an enforced hill-climbing algorithm. In: Foundations of intelligent systems. Springer, pp 216–227

Karaboga D, Bastruk B (2007) A powerful and efficient algorithm for numeric function optimization: artificial bee colony (ABC) algorithm. J Glob Optim 39(3):459–471

Kennedy J (2010) Particle swarm optimization encyclopedia of machine learning. Springer, pp 760–766

Kifah S, Abdullah S (2015) An adaptive non-linear great deluge algorithm for the patient-admission problem. Inf Sci 295:573–585

Kirkpatrick S, Gelatt CD, Vecchi MP (1983) Optimization by simulated annealing. Science 220(4598):671–680

Kothari DP (2012) Power system optimization. Paper presented at the 2nd international conference on computational intelligence and signal processing (CISP)

Kumbharana N, Pandey GM (2013) A comparative study of ACO, GA and SA for solving travelling salesman problem. Int J Soc Appl Comput Sci 2(2):224–228

Ledesma S, Aviña G, Sanchez R (2008) Practical considerations for simulated annealing implementation. Simul Anneal 20:401–420

Li X-G, Wei X (2008) An improved genetic algorithm-simulated annealing hybrid algorithm for the optimization of multiple reservoirs. Water Resour Manage 22(8):1031–1049

Li L, Chu W, Langford J, Schapire RE (2010) A contextual-bandit approach to personalized news article recommendation. Paper presented at the proceedings of the 19th international conference on world wide web

Li M, Du W, Nian F (2014) An adaptive particle swarm optimization algorithm based on directed weighted complex network. Math Probl Eng 2014(2014):1–7

Mahdavi M, Fesanghary M, Damangir E (2007) An improved harmony search algorithm for solving optimization problems. Appl Math Comput 188(2):1567–1579

Malek M, Guruswamy M, Pandya M, Owens H (1989) Serial and parallel simulated annealing and tabu search algorithms for the traveling salesman problem. Ann Oper Res 21(1):59–84

Manjarres D, Landa-Torres I, Gil-Lopez S, Del Ser J, Bilbao MN, Salcedo-Sanz S, Geem ZW (2013) A survey on applications of the harmony search algorithm. Eng Appl Artif Intell 26(8):1818–1831

Mcmullan P (2007) An extended implementation of the great deluge algorithm for course timetabling. In: Computational science-ICCS 2007. Springer, pp 538–545

Minton S, Johnston MD, Philips AB, Laird P (1992) Minimizing conflicts: a heuristic repair method for constraint satisfaction and scheduling problems. Artif Intell 58(1–3):161–205

Mirjalili S, Mirjalili SM, Lewis A (2014a) Let a biogeography-based optimizer train your multi-layer perceptron. Inf Sci 269:188–209

Mirjalili S, Mirjalili SM, Lewis A (2014b) Grey wolf optimizer. Adv Eng Softw 69:46–61

Mohapatra P, Das S (2013) Stock market prediction using bio-inspired computing: a survey. Int J Eng Sci Technol 5(4):739–746

Nabeel R (2010) Hybrid genetic algorithms with great deluge for course timetabling. Int J Comput Sci Netw Secur 10:283–288

Nahas N, Kadi DA, El Fath MN (2010) Iterated great deluge for the dynamic facility layout problem. CIRRELT

Nozohour-Leilabady B, Fazelabdolabadi B (2016) On the application of artificial bee colony (ABC) algorithm for optimization of well placements in fractured reservoirs; efficiency comparison with the particle swarm optimization (PSO) methodology. Petroleum 2(1):79–89

Othman ZA, Theng LM, Zainudin S, Sarim HM (2013) Great Deluge algorithm feature selection for network intrusion detection. J Appl Sci Agric 8(4):322–330

Osaba E, Yang X-S, Diaz F, Lopez-Garcia P, Carballedo R (2016) An improved discrete bat algorithm for symmetric and asymmetric traveling salesman problems. Eng Appl Artif Intell 48:59–71

Osman IH, Kelly JP (2012) Meta-heuristics: theory and applications. Springer Science & Business Media

Özcan E, Misir M, Ochoa G, Burke EK (2012) A reinforcement learning: great-deluge hyper-heuristic. In: Modeling, analysis, and applications in metaheuristic computing: advancements and trends, p 34

Park C, Pan J, Manocha D (2013) Real-time optimization-based planning in dynamic environments using GPUs. Paper presented at the IEEE international conference on robotics and automation (ICRA)

Pierre DM, Zakaria N, Pal AJ (2011) Master-slave parallel vector-evaluated genetic algorithm for unmanned aerial vehicle's path planning. Paper presented at the 11th international conference on hybrid intelligent systems (HIS)

Prakasam A, Savarimuthu N (2015) Metaheuristic algorithms and probabilistic behaviour: a comprehensive analysis of ant colony optimization and its variants. Artif Intell Rev 1–34

Rani D, Jain SK, Srivastava DK, Perumal M (2012) Genetic algorithms and their applications to water resources systems. In: Metaheuristics in water, geotechnical and transport engineering, p 43

Razali NM, Geraghty J (2011) Genetic algorithm performance with different selection strategies in solving TSP. Paper presented at the proceedings of the world congress on engineering

Safari SA (2015) Wildlife and big 5. Accessed 12 Dec 2018. http://sasafari.com/wildlife-and-the-big-five/

Seidy EE (2016) A new particle swarm optimization based stock market prediction technique. Int J Adv Comput Sci Appl 7(4):322–327

Selman B, Gomes CP (2006) Hill-climbing search. Encyclopedia of cognitive science, 81(333–335):10

Senthilnath J, Das V, Omkar S, Mani V (2012) Clustering using Lévy flight cuckoo search. Paper presented at the BIC-TA (2)

Sharma A, Sharma A, Panigrahi B, Kiran D, Kumar R (2016) Ageist spider monkey optimization algorithm. Swarm Evol Comput

Tambouratzis T (1997) A simulated annealing artificial neural network implementation of the N-queens problem. Int J Intell Syst 12(10):739–751

Van Laarhoven PJ, Aarts EH, Lenstra JK (1992) Job shop scheduling by simulated annealing. Oper Res 40(1):113–125

Wen TH, Gasic M, Kim D, Mrksic N, Su PH, Vandyke D, Young S (2015). Stochastic language generation in dialogue using recurrent neural networks with convolutional sentence reranking. arXiv preprint arXiv:1508.01755

Wong L, Moin NH (2015) Enhanced ant colony optimization for inventory routing problem. Paper presented at the 22nd national symposium on mathematical sciences (SKSM22): strengthening research and collaboration of mathematical sciences in Malaysia

Xi B, Liu Z, Raghavachari M, Xia CH, Zhang L (2004) A smart hill-climbing algorithm for application server configuration. Paper presented at the proceedings of the 13th international conference on world wide web

Yang X-S (2011) Review of meta-heuristics and generalised evolutionary walk algorithm. Int J Bio-Inspir Comput 3(2):77–84

Yang C-T, Sung T-C, Weng W-C (2006) An improved tabu search approach with mixed objective function for one-dimensional cutting stock problems. Adv Eng Softw 37(8):502–513

Yang X-S, Deb S, Fong S (2014) Metaheuristic algorithms: optimal balance of intensification and diversification. Appl Math Inf Sci 8(3):977

Chapter 4
A Brief Review of Meta-heuristics

Abstract This chapter presents a detailed overview of Swarm Intelligence methods, which are inspired by the collective behaviour of natural systems such as birds, ants, bees, and wolves. It explores how these algorithms work using simple agents that cooperate to solve complex problems without centralized control. The chapter highlights key swarm-based algorithms, including Particle Swarm Optimization (PSO), Ant Colony Optimization (ACO), Genetic Algorithm (GA), Evolutionary Strategy (ES), Probability-Based Incremental Learning (PBIL), and Biogeography-Based Optimization (BBO). Each algorithm is explained with its working principle, pseudocode, applications, strengths, and limitations, particularly in the context of stock market prediction. These methods show potential in addressing real-world data challenges by enabling better learning, optimization, and decision-making. The chapter concludes with a discussion on how BBO and other swarm-based algorithms can be used with neural networks for improved predictive accuracy, setting the stage for the proposed ensemble model in later chapters.

Keywords Swarm Intelligence · Particle Swarm Optimization · Ant Colony Optimization · Genetic Algorithm · Evolutionary Strategy (ES) algorithm · Probability based incremental learning (PBIL) · Biogeography-based optimization (BBO)

4.1 Introduction

A number of researchers investigated and specified a new discipline named Swarm Intelligence that consists of a simple set of mobile agents to solve essential issues collectively by direct or indirect communications (Binitha and Sathya 2012; Kennedy et al. 2001; Mahale and Chavan 2012). Swarm Intelligence utilizes basic rules consisting emergence of intelligent behavior for secret single agent to treat a group of individual natural and artificial systems for organizing by decentralization. Ant Colony Optimization (ACO) and Particle Swarm Optimization (PSO) are two of the

D. Das et al., *Optimization Algorithms in Machine Learning*, Engineering Optimization: Methods and Applications, https://doi.org/10.1007/978-981-96-3849-9_4

commonly used Swarm Intelligence algorithms. As Grey Wolf Optimizer (GWO) is the main algorithm in this book, it will be investigated further in detail.

4.2 Swarm-Based Approaches

Swarm Intelligence consists of combined social interactions of creatures that implement group's cooperative intelligence for simple agents like ants, animals, plants, and other elements of ecosystem depending on the real-life behavior (Pandiri and Singh 2015). Swarm-based approaches comprise various features like:

- Swarm-based approaches are population-based and use multi-agents for searching.
- The population agents are homogeneous.
- The outcomes of the system yield from individual interactions with each other in the environment.
- The movement of the individual agents is mobile and chaotic.
- The control structure is decentralized, where each iteration is performed by the action of individual leader (Parpinelli and Lopes 2011).

Ant Colony Optimization (ACO), Particle Swarm Optimization (PSO), Bee Colony Optimization (BCO), and Firefly Algorithm (FA) are some of the prominent and frequently applied Swarm-based approaches. The mentioned common Swarm-based approaches may be applied for stock prediction. Enormous amount of information is processed and stored every day in stock markets worldwide. However, it is not always possible to make an appropriate decision about the stock investment using this information. Sometimes, it may not be even possible to receive the desired return through the stock investment from this vast amount of information applying various predictive models.

Stock market remains the best investment alternatives for few decades despite being unpredictable and uncertain. Prediction of stock price is extremely complicated due to the nonlinear form of stock data. As the economic condition of a country greatly depends on stock market, researchers are investigating endlessly to determine the best predictive model for stock market. Prediction of stock market is significant in finance and is gaining more attention of the researchers, due to the fact that the investors may be better guided through successful prediction of the stock price.

Exploring stock data needs to build a predictive or descriptive model such that hidden information lies in data can be unfolded. Consequently, building a predictive model from the stock data is a complicated task. Information from large databases can be extracted through a well-known technology called data mining that facilitates the organizations to retrieve the vital information from data repositories (Witten et al. 2016).

From the existing models and recently developed algorithms, extracting the best subset of features that helps in accurately identifying the labeled action from stock market massive amount of data 2^N subsets is not an easy task and tends to be non-polynomial complex problem during the raise of searching space (Chandrashekar and Sahin 2014). Considering the limitations and prospects, research should be focused to develop an algorithm for stock prediction that is applicable not only for a single market but also generalized for various stock markets.

However, applying machine learning model does not guarantee a good accuracy, besides the accuracy of machine learning models similar with data mining as well as neural network may be affected by numerous factors (Negnevitsky 2005). Moreover, the selection of input parameters may result in inconsistent output.

4.3 Particle Swarm Optimization

Particle Swarm Optimization (PSO) is one of the common choices for solving complex and intricate problems that cannot be solved by traditional methods. Stock prediction issue can be well addressed through PSO that facilitates maximizing profit and minimizing risk. The application of PSO is presented by Cura (2009) and Particle Swarm with Center of Mass (PSOCoM) to move the particles to the best-predicted position is proposed by Seidy (2016) that can train the adaptive linear combiner to form a stock market prediction (Seidy 2016).

PSO has been combined with other models to propose ensemble model for stock prediction in recent work (Khajavi and Amiri 2017; Seidy 2016). Many optimization problems have been addressed through PSO. The natural process of swarm behaviors such as bird and fish swarm for searching food is mimicked by PSO. The local search by individual experience with the global search by neighboring experience can be balanced by PSO.

The pseudocode of PSO algorithms is indicated in Fig. 4.1. PSO applies swarm (population) of particles (individuals), which can be moved to the search space over numerous iterations. Each particle indicates a candidate solution for the problem, which is also considered as a point in M-dimensional space. The status of the particle is portrayed by its position and velocity.

```
Begin
While terminating condition is not reached do
    For i = 1 to number of particles do
        Calculate the fitness function f
        Update personal best and global best of each particle
        Update velocity of the particle using Equation, V(t+1) = wV(t) +
            c1 ×rand () × (Xpbest - X(t)) + c2 ×rand () × (Xgbest - X(t))
        Update the position of the particle using Equation, X (t+1) = X(t) + V(t+1)
    End For
End While
End
```

Fig. 4.1 PSO algorithms pseudocode

PSO is accomplished by adjusting a swarm of random particles where particle flying along the direction that will be balanced through local best (position of one particle) and global best (ever found by all particles). The particle is updated in each iteration by two best values or fitness, namely *pBest* (local best) and *gBest* (global best).

PSO has been applied for nonlinear function optimization, pharmaceutical and biomedical applications, communications and combinatorial optimization problems successfully (Poli 2007). The critical assessment of PSO algorithms confirms that PSO algorithms are better than Ant Colony Optimization (ACO) as PSO algorithms are simple for implementation and a small number of parameters are needed to adjust (Pereira 2011).

Also, PSO provides greater diversification. Additionally, the memory capacity of PSO is efficient and better than GA. But, PSO does not utilize evolution operators such as mutation, crossover, inversion, and selection. PSO is similar to ES, GA, and GP in terms of initialization and updating generations.

Although PSO algorithms are efficient for searching both continuous and multi-modal, however, there are some limitations of the PSO algorithm such as error rate is high in some situations and the performance may not be that well after the implementation of internal and external factors (Khajavi and Amiri 2017). Moreover, utilization of multiple parameters by PSO algorithms may affect the efficiency and speed (Tanweer et al. 2015). In addition, limitations such as easily falling into local optimum in high-dimensional space and having a low convergence rate in the iterative process motivates the researcher to either use different algorithms or improve PSO (Li et al. 2014).

4.4 Ant Colony Optimization

Ant Colony Optimization (ACO) is an evolutionary algorithm that mimics the behavior of Ant Colony which can solve the complicated combination optimization problems, i.e. Traveling Salesman Problem (TSP). Initially, an ant locates the food source and returns to the nest. Ants observe four possible ways extensively, but the runway is consolidated in a way that the route is not less attractive than shortest route. Though ants lose their trail pheromones, they follow the shortest route (Mohapatra and Das 2013). Yang et al. (2014) applied combinatorial model to predict short-term electricity price of New South Wales in Australia by ACO algorithm based on the generalized autoregressive conditional heteroskedasticity (GARCH) model and SVM. The forecasting accuracy is improved through their model (Yang et al. 2014).

ACO has been integrated with other models to form the ensemble approach for stock prediction in recent investigations (Cai et al. 2015; Yang et al. 2014). ACO can be applied to determine an appropriate partition of stock data through engaging ants for searching. The balance between exploration and exploitation can be made through ACO algorithm where pheromone intensification of paths and exploitation is

```
Begin
1: Set parameters, initialize pheromone trails
2: While terminating condition is not reached do
3:    ConstructAntSolutions applying pheromone trail
4:    Update Pheromones
5: End While
End
```

Fig. 4.2 ACO algorithms pseudocode

the main focus (Cai et al. 2015). Better predication accuracy can be availed through ACO integrated with other model (Cai et al. 2015).

The pseudocode of ACO algorithm is indicated in Fig. 4.2. In ACO algorithm, ConstructAntSolutions is a partial solution extended by adding an edge based on stochastic and pheromone considerations. Update pheromone is a process to increase pheromone of good solutions, decrease that of bad solutions, which is also known as pheromone evaporation.

ACO has some advantages like inherent parallelism, efficiency for TSP and similar problems and suitable for dynamic application. ACO can be implemented to solve various issues like machine learning problems, network problems, stochastic optimization problems, and Traveling Salesman Problems (Stützle et al. 2011). ACO can be hybridized or ensemble with other algorithms to form robust and efficient algorithms. Moreover, ACO is very effective for distributed environment. However, loss of diversity and increased chance of premature convergence are some of the limitations of ACO algorithm (Cai et al. 2015).

In addition, ACO algorithms utilize multiple parameters such as pheromone quantity, pheromone update rule, evaporation rate, and pheromone reinforcement rate, which need to be tuned properly. In addition, difficulty in theoretical analysis and changing the probability distribution after each course of iteration has motivated the researchers to investigate for suitable algorithms (Mohapatra and Das 2013).

4.5 Genetic Algorithm

Genetic Algorithm (GA) is a population-based meta-heuristic evolutionary algorithm that mimics biological evolution, and it can solve constrained and unconstrained optimization problem through natural selection process. The population of individual solution is repetitively modified through GA algorithm. An individual from current population is randomly chosen by the algorithm, which is used as a parent to generate children for next generation. The successive generations finally produces the optimal solution through the evolution of the population. Delnavaz (2014) applied GA and fuzzy-neural network algorithm to predict the stock price for Tehran Stock Market. The result through the combinatorial algorithms is encouraging (Delnavaz 2014).

GA has been integrated with other models to form the ensemble approach for stock prediction in recent investigations (Göçken et al. 2016; Delnavaz 2014). GA

```
Begin
1: Generate an initial random population
2: Evaluate each candidate solution
3: While terminating condition is not reached do
4:    Select individuals for the next generation
5:    Recombine pairs of parents
6:    Mutate the resulting offspring
7:    Evaluate each candidate solution
8: End While
End
```

Fig. 4.3 GA algorithm pseudocode

can be applied to overcome the limitation of input variable selection and also it is potential for search and optimization problem. During evolution, GA can generate new and better population among different species. GA is capable to exploit the unknown search space through the collected information (Göçken et al. 2016).

The pseudocode of GA for stock data classification is directed in Fig. 4.3. Selection, crossover, and mutation are the three basic operators of GA. However, GA can be extended for better performance through the adjustment of elitism (best individuals in a population can be propagated to the next generation) or random immigrants (worst individuals in a population can be replaced by random one).

Nevertheless, there are some limitations that persist with GA algorithm such as hidden layer has to remain fixed due to time-consuming training, transfer, and training function need to be fixed as the combination of both may deteriorate the quality (Göçken et al. 2016). Moreover, difficulty in identifying the fitness function because of the occurrence of premature convergence, complexity in choosing population because of mutation rate, crossover rate, time-consuming decoding, and fitness evaluation may be experienced with GA. Hence, the limitations of GA divert the researcher to investigate for other good algorithms (Razali and Geraghty 2011).

4.6 Evolutionary Strategy

Evolutionary Strategy (ES) algorithm is capable to optimize and search the space through the simulation of the genetic evolutionary process proposed by Darwin's theory consisting of selection, mutation, recombination, and reproduction. The algorithm measures the performance through the application of fitness function and efforts the evolution place in better search space regions.

ES algorithm has been integrated with other models to form the ensemble approach for stock prediction in recent investigations (Hu et al. 2015; Bliss et al. 2014; Bisoi and Dash 2014). Hu et al. (2015) investigated and proposed that EA algorithm can be applied for rule discovery in stock algorithm trading (AT) (Hu et al. 2015). Bliss et al. (2014) applied EA algorithm to predict future links in social networks by applying the Covariance Matrix Adaptation Evolution Strategy (CMA-ES) to optimize weights

```
Begin
Create a population of individuals
While terminating condition is not reached do
    Evaluate the fitness of each of the individuals
    Select the best individuals for reproduction to form the next generation
    Perform reproduction on the parent solutions to form new child solutions
    Perform mutations on the child solutions
End While
End
```

Fig. 4.4 ES algorithm pseudocode

(Bliss et al. 2014). Bisoi and Dash (2014) proposed the use of combinatorial evolutionary dynamic neural network for stock market trend analysis and prediction using unscented Kalman filter.

The pseudocode of ES is indicated in Fig. 4.4. In ES, a population of individual is created so that it represents the solution of a problem. The process is considered as a solution of some problems, and it is similar to genes in natural evolution. The fitness of each gene is assessed to measure the capability in solving problems. The best gene is chosen at this stage to reproduce for next generation, and new child solution is formed from the parent through reproduction. The mutation on the child solution is executed, and the process is iterated till suitable solution has been achieved.

However, ES algorithm has some limitations such as finding optimal solutions in a finite amount of time is not guaranteed, parameter tuning mostly by trial-and-error, and population approach may be expensive in terms of other meta-heuristic algorithms, influences the researchers to look for alternate solution (Bisoi and Dash 2014).

4.7 Probability-Based Incremental Learning

Probability-based incremental learning (PBIL) is an evolutionary optimization algorithm, which is able to create the real-valued probability vector for the object of the algorithm, and it can produce high evaluation solution vectors with high probability through sampling (Baluja 1994). PBIL algorithm is a better algorithm for solving real-world problems than GA and hill-climbing, and it is formed through the generalization of GA to preserve the statistics of population produced by GA.

PBIL algorithm has been integrated to form the ensemble approach for stock prediction (Monteiro et al. 2018; Ali et al. 2014). Monteiro et al. (2018) investigated the application of probability-based ensemble model to predict for a day ahead Iberian Electricity Market. Ali et al. (2014) applied PBIL algorithm for determining the Egyptian stock market trend through the enhancement of the performance of multi-layer perceptron and achieved better result. Numerous researches have been concentrated on incremental learning than selective learning.

```
Begin
1: Initialize the probability vector P(i) = 0.5
2: While terminating condition is not reached do
3:   M = generate samples from probability vector P
4:   Evaluate samples(M)
5:   B = select best solutions from(M)
6:   P(i) = (1-α) * P(i) + α * B(i)
7: End While
End
```

Fig. 4.5 PBIL algorithm pseudocode

The pseudocode for PBIL algorithm is indicated through Fig. 4.5. PBIL uses an initial probability vector initialized to 0.5 for every entry. The reason for choosing 0.5 is that the probability of generating 1 or 0 for each course of iteration is equal (Monteiro et al. 2018) but the values will be updated through learning as the search continues.

However, some of the drawbacks of PBIL algorithm such as PBIL algorithm depends on the inversion of information matrix, PBIL can only converge to local optima though in case of unimodal functions PBIL can converge to the global optimum, and PBIL uses single probability vector which may have less expressive power (Monteiro et al. 2018) can be addressed either by implementing other optimization algorithm or improving PBIL (Ali et al. 2014).

4.8 Biogeography-Based Optimization

The optimization of neural network can be performed through the application of biogeography-based optimization (BBO) (Mirjalili et al. 2014) in training MLPs. BBO is an evolutionary algorithm that applies evolutionary mechanisms to each individual in a population. BBO can provide more flexible training procedures compared to others for the search space of MLP that is changeable for different datasets. It tends to outperform GA due to applying various evolutionary operators.

The pseudocode of BBO is indicated in Fig. 4.6. BBO algorithm will initially outline the island modification probability, mutation probability, and elitism parameter and initialize the population. The immigration rate and emigration rate will be calculated for each island, provided that the solution will be considered as good if it has high emigration rates and low immigration rates. Otherwise, if it has low emigration rates and high immigration rates then the solution will be treated as bad.

Here, the immigration islands will be chosen based on the immigration rates probabilistically and roulette wheel selection will be used based on the emigration rates to select the emigrating islands. Then, randomly selected Suitability Index Variables (SIVs) will be migrated based on the selected islands where the migration will take place randomly. BBO performs mutation based on the mutation probability

```
Begin
Generate an initial random population
While terminating condition is not reached do
  Calculate the immigration rate and emigration rate for each island
  Choose the immigration islands based on the immigration rates probabilistically
  Migrate randomly selected Suitability Index Variables (SIVs) based on the
  selected islands
  Perform mutation based on the mutation probability for each island
  Probabilistically
  Calculate the fitness of each individual island
End While
End
```

Fig. 4.6 BBO algorithm pseudocode

for each island probabilistically. Finally, fitness of each individual island will be calculated and the process continues until the target is achieved.

BBO is a meta-heuristic algorithm that applies evolutionary mechanisms to each individual in a population. BBO can provide more flexible training procedures compared to others for the search space of MLP that is changeable for different datasets. It tends to outperform GA due to applying various evolutionary operators (Mirjalili et al. 2014). Usually, heuristic algorithms are employed for solving a particular problem by determining a combination of weights and biases that provide the minimum error for an MLP. The architecture does not change during the learning process in this method. For minimizing the overall error of MLP, the training algorithm needs to discover proper values for all connection weights and biases.

Generally speaking, there are three methods of using a heuristic algorithm for training MLPs. Firstly, heuristic algorithms are utilized for searching. Secondly, heuristic algorithms are employed to find a proper architecture for an MLP in a particular problem. The last method is to use a heuristic algorithm to tune the parameters of a gradient-based learning algorithm, such as the learning rate and momentum. The weights and biases are encoded using vector to train an MLP. The encoding is easier in this way though the decoding is a bit complicated. This method is used often for simple neural network structure, and it is appropriate for the problem, which can't deal with complex MLP structure (Haykin 1994).

BBO algorithm can be integrated to form ensemble approach for better classification and solving prediction problem. Moreover, BBO has much scope to grow as this research community is quite young. Significant challenging tasks can be addressed through BBO by exploring new approach (Zhang et al. 2016; Mirjalili et al. 2014). However, BBO algorithm may have some limitations such as poor in exploiting the solutions, no provision for selecting the best members from each generation and a habitat doesn't consider its resultant fitness while immigrating the features may result in the generation of many infeasible solutions. The extension of BBO and ensemble with other models may be investigated to address the limitations (Ammu et al. 2013).

4.9 Summary

In this chapter, a brief review of meta-heuristics methods and their algorithms was presented and discussed. Their main features as well as limitations were also investigated in this chapter. Toward the end of this chapter, a quite similar ensemble model to our proposed model in this book, that is BBO, and neural network was discussed and its limitations highlighted also. To this end, the following chapters will present further literature on the related NN training algorithms.

References

Ali BAS, Ahmed ABED, Muhammad AED (2014) Incremental learning approach for enhancing the performance of multi-layer perceptron for determining the stock trend. Int J Sci Basic Appl Res (IJSBAR) 16(1):15–23

Ammu PK, Sivakumar KC, Rejimoan R (2013) Biogeography-based optimization—a survey. Int J Electron Comput Sci Eng 2(1):154–160

Baluja S (1994) Population-based incremental learning: a method for integrating genetic search based function optimization and competitive learning. In: Population based incremental learning, CMU-CS-94-163, pp 1–41

Binitha S, Sathya SS (2012) A survey of bio inspired optimization algorithms. Int J Soft Comput Eng 2(2):137–151

Bisoi R, Dash PK (2014) A hybrid evolutionary dynamic neural network for stock market trend analysis and prediction using unscented Kalman filter. Appl Soft Comput 19:41–56

Bliss CA, Frank MR, Danforth CM, Dodds PS (2014) An evolutionary algorithm approach to link prediction in dynamic social networks. J Comput Sci 5(5):750–764

Cai Q, Zhang D, Zheng W, Leung SC (2015) A new fuzzy time series forecasting model combined with ant colony optimization and auto-regression. Knowl-Based Syst 74:61–68

Chandrashekar G, Sahin F (2014) A survey on feature selection methods. Comput Electr Eng 40(1):16–28

Cura T (2009) Particle swarm optimization approach to portfolio optimization. Nonlinear Anal Real World Appl 10(4):2396–2406

Delnavaz B (2014) Forecasting of the stock price index by using fuzzy-neural network and genetic algorithms. J Appl Sci Agric 9:109–117

Göçken M, Özçalıcı M, Boru A, Dosdoğru AT (2016) Integrating metaheuristics and artificial neural networks for improved stock price prediction. Expert Syst Appl 44:320–331

Haykin S (1994) Neural networks, a comprehensive foundation (0023527617). Macmillan, USA

Hu Y, Liu K, Zhang X, Su L, Ngai E, Liu M (2015) Application of evolutionary computation for rule discovery in stock algorithmic trading: a literature review. Appl Soft Comput 36:534–551

Kennedy J, Kennedy JF, Eberhart RC, Shi Y (2001) Swarm intelligence. Morgan Kaufmann

Khajavi S, Amiri FS (2017) Prediction of stock price using particle swarm optimization algorithm and Box–Jenkins time series. Int J Financ Manag Account 2(7):25–31

Li M, Du W, Nian F (2014) An adaptive particle swarm optimization algorithm based on directed weighted complex network. Math Probl Eng 2014(2014):1–7

Mahale RA, Chavan S (2012) A survey: evolutionary and swarm based bio-inspired optimization algorithms. Int J Sci Res Publ 2(12):1–6

Mirjalili S, Mirjalili SM, Lewis A (2014) Let a biogeography-based optimizer train your multi-layer perceptron. Inf Sci 269:188–209

Mohapatra P, Das S (2013) Stock market prediction using bio-inspired computing: a survey. Int J Eng Sci Technol 5(4):739–746

Monteiro C, Ramirez-Rosado IJ, Fernandez-Jimenez LA, Ribeiro M (2018) New probabilistic price forecasting models: application to the Iberian electricity market. Int J Electr Power Energy Syst 103:483–496

Negnevitsky M (2005) Artificial intelligence: a guide to intelligent systems. Pearson Education, USA

Pandiri V, Singh A (2015) Swarm intelligence approaches for multidepot salesmen problems with load balancing. Appl Intell 1–13

Parpinelli RS, Lopes HS (2011) New inspirations in swarm intelligence: a survey. Int J Bio-Inspir Comput 3(1):1–16

Pereira G (2011) Particle swarm optimization. INESCID and Instiuto Superior Tecnico, Porto. Accessed 13 Dec 2018. Available at: https://pdfs.semanticscholar.org/0033/9e89348fdda5b01 9591de93f0c6bf229bf4b.pdf.93f0c6bf229bf4b.pdf

Poli R (2007) An analysis of publications on particle swarm optimization applications. Department of Computer Science, University of Essex, Essex, UK

Razali NM, Geraghty J (2011) Genetic algorithm performance with different selection strategies in solving TSP. Paper presented at the proceedings of the world congress on engineering

Seidy EE (2016) A new particle swarm optimization based stock market prediction technique. Int J Adv Comput Sci Appl 7(4):322–327

Stützle T, López-Ibáñez M, Dorigo M (2011) A concise overview of applications of ant colony optimization. In: Wiley encyclopedia of operations research and management science

Tanweer M, Suresh S, Sundarajan N (2015) Improved SRPSO algorithm for solving CEC 2015 computationally expensive numerical optimization problems. Paper presented at the 2015 IEEE congress on evolutionary computation (CEC)

Witten IH, Frank E, Hall MA, Pal CJ (2016) Data mining: practical machine learning tools and techniques. Morgan Kaufmann, USA

Yang X-S, Deb S, Fong S (2014) Metaheuristic algorithms: optimal balance of intensification and diversification. Appl Math Inf Sci 8(3):977

Zhang Y, Phillips P, Wang S, Ji G, Yang J, Wu J (2016) Fruit classification by biogeography-based optimization and feed-forward neural network. Expert Syst 33(3):239–253

Chapter 5
Artificial Neural Networks: Structure and Learning

Abstract This chapter presents the details about Artificial Neural Network (ANN) that controls human intelligence, its structure and learning techniques. It outlines how an ANN substantially contributes towards neural network's learning as well as how node, weight, and layers can be adjusted to construct an appropriate ANN for a given scenario. Various types of ANN is discussed in this chapter along with its application to solve time series prediction. The chapter concludes with the direction of appropriate learning algorithm based on the investigation.

Keywords Artificial Neural Network (ANN) · Meta-heuristic algorithms · Training neural network · Neuron · Bias · Activation function · Grey Wolf Optimizer (GWO) · Ensemble model

5.1 Introduction

The architecture of human brain consists parallel neurons network, which can enormously control human intelligence. Essentially, an appropriate outline of an ANN which is also known as Neural Networks (NNs) can substantially contributes to its learning. Node, weight, and layers can be adjusted to construct an appropriate ANN based on a problem at hand. ANN can be single layer or multi-layer. However, single layer consists of one input and an output layer which is appropriate for solving linear problems.

On the other hand, a multi-layer consists of input layer, output layer, and one or more hidden layers that can solve nonlinear problems. ANN and its improvement for time series prediction have been investigated recently by many researchers for numerous investigations (Wanto et al. 2017; Lahmiri and Boukadoum 2015; Balabanov et al. 2011). Back-propagation algorithm is used before to train multi-layer neural network but appropriate learning algorithm can significantly improve the performance of neural network for pattern recognition, prediction, and many other diverse applications.

D. Das et al., *Optimization Algorithms in Machine Learning*, Engineering Optimization: Methods and Applications, https://doi.org/10.1007/978-981-96-3849-9_5

5.2 Neural Networks

Artificial Neural Network or ANN is one of the best technologies that have been extensively applied to many different fields over the last few decades. It is one of the most widespread and key methods of machine learning (ML) nowadays. ANN is a kind of mathematical model motivated by similar structure and function of human brain. Artificial neurons are the building blocks of ANN where each neuron is connected together by weights and learning of ANN, which is the most important criterion for the better performance of ANN, takes place by adjusting those weights. The nonlinearity of the ANN model is handled by a special function called activation function, where another important factor called Bias (similar to y-intercept in a line equation) is implemented to determine the shifting of the activation function either toward left or right. The learning of ANN can be performed by different learning techniques such as Supervised, Unsupervised, or Reinforcement, where supervised learning takes place with the help of labeled data, unsupervised learning occurs with unlabeled data and reinforcement learning materializes through an agent without any specific dataset. The greatest flexibility of ANN is application of learning function to adapt it to solve various problems like biological, medical, industrial, financial, stock market, software engineering, environmental, economical, social applications, and many other similar areas.

Neuron, an abstraction of the biological neuron, is the basic structural unit of the Neural Networks (NNs) and flow of information can affect the structure of NN. Thus, input and output can change the NN. Moreover, nonlinear statistical data can be processed by this NN and hence complex relationship can be built between input and output. It is possible to build different patterns through NN. As indicated in Fig. 5.1, the neural network functions similar to the human brain connected internally in an appropriate manner, whereas the connection in brain is modeled through neurons and dendrites that is modeled through silicon and wires in NN.

As demonstrated in Fig. 5.1, the fundamental processing element of a neural network is a neuron. Human brain has approximately 86 billion neurons, whereas ant brain has 250,000 neurons. The input and output units are connected in a neural network such a way that each connection has associated weight. It makes the neural network as a connectionist learning. The learning of a neural network takes place by adjusting the weights, so that it can correctly classify the training data one end and classify the unknown data after testing phase another end. The most time-consuming part of neural network is the learning. Another exciting feature of neural network is that it has better tolerance to noisy and incomplete data.

Thus, neural network consists of multiple nodes that mimic the similar behavior of human brain or biological neurons. As indicated in Fig. 5.3, each node communicates with each other as they are connected by links. Consequently, nodes accept input data, perform simple operations, pass these operations toward other neurons, and produce output through activation function. Here, every link of a neural network contains an associated weight and it has the capability to learn by altering those weights.

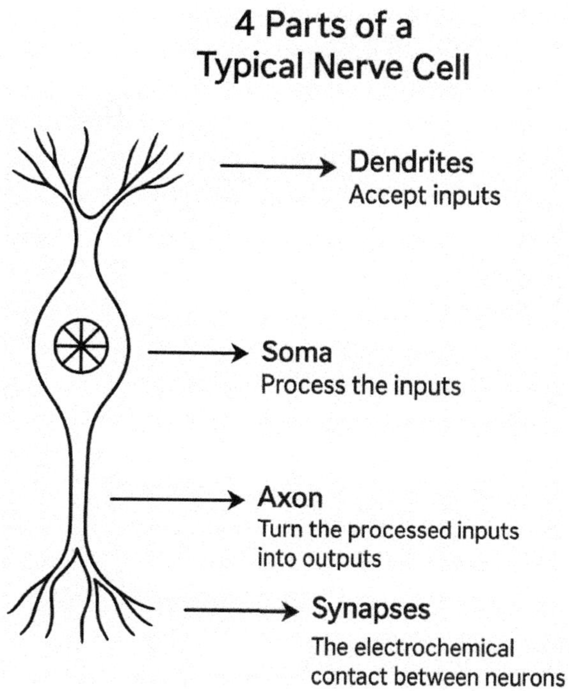

Fig. 5.1 Basic structure of neural network

As indicated in Fig. 5.2, the processing done by a neuron of ANN can be denoted as:

$$\text{output} = \text{sum}\,(\text{weights} * \text{inputs}) + \text{bias}$$

Neural network, a widely leveraged complex nonlinear problem solver, can have two main types, namely FeedForward neural network and Feedback neural network. As shown in Fig. 5.3, in case of FeedForward neural network, the information can

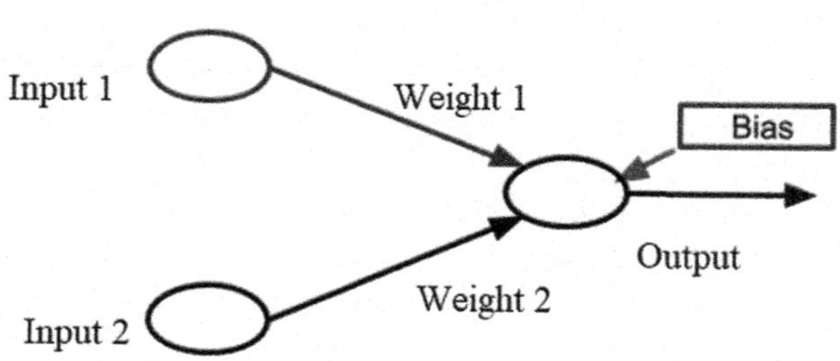

Output = Weight 1 * Input 1 + Weight 2 * Input 2 + Bias

Fig. 5.2 Processing by a neuron

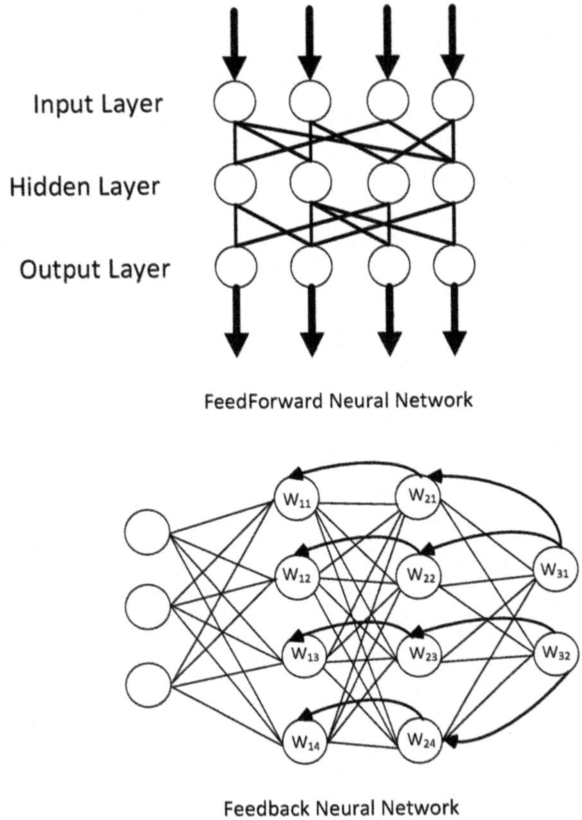

Fig. 5.3 FeedForward and feedback neural network

flow only in one direction. Thus, the sender will not receive anything and it does not contain any feedback loops. This type of neural network can be applied for pattern recognition as it comprises fixed input and output. On the other hand, feedback neural network comprises feedback loops that can be applied in content addressable memories, signal processing, optimal computation, etc. However, the optimization of FeedForward neural network remains one of the most popular researches worldwide over the years, where the optimization can be performed by optimizing the weights, architecture of the network, number of nodes, learning parameter, environment, and so on.

The neural network needs to be trained to generate the output or target much closer to desired one. Berry and Linoff (1997) defined the 'Training' as a process of producing, finding or setting the weights in a neural network to produce good prediction result. Numerous algorithms are offered by neural network for the purpose of training but back-propagation is the widely accepted one for training multi-layer perceptron network among the available options (Wu and Coggeshall 2012) due to its ability to faster convergence and mathematical compliance. Levenberg–Marquardt and Gradient Decent are extensively used for training through back-propagation;

however, they are computationally expensive to support large neural networks especially, when the network to be trained consists substantial amount of adaptive weights (Schmidhuber 2015). Figure 5.4 indicates the back-propagation algorithm.

The basic structure of a multi-layer back-propagation neural network is indicated in Fig. 5.5.

Another effective ANN-based architecture that has shown great potential in solving complex prediction problems over the years is Deep Learning (DL). In this big data era with huge storage capability, better management, quicker updating, speedy data collection and availability of GPU contribute to process huge dataset. Therefore, it is the best time to grow and implement Deep Learning for solving prediction or other similar problems as it can significantly reduce the computation time as well as facilitate faster convergence. DL is able to excerpt high-level features from raw input using multiple layers. Similar to ANN, DL has been applied to solve the problems like classification, pattern recognition, nonlinear system identification, hand-written digit recognition, speech recognition, medical image processing, and

Input: *D*, a dataset consisting of the training tuples and their associated target values
L, the learning rate Network, a multilayer feed forward network
Output: A trained network
Begin
Initialize all network weights and biases
While terminating condition is not satisfied do
  For each training tuple *X* in *D*
    For each input layer unit *j*
      Output of an input unit, O_j = actual output value, I_j
    End For
    For each hidden or output layer unit *j*
      Compute the net input of unit *j* with respect to the previous layer, *i*
      $I_j = \sum_i w_{ij} O_i + \Theta_j$
      Compute the unit of each unit *j*, $O_j = \frac{1}{1+e^{-I_j}}$
    End For
    For each unit j in the output layer
      Computer the error, $Err_j = O_j(1- O_j)(T_j - O_j)$
    End For
    For each unit j in the hidden layer,
      Compute the error with respect to the next higher layer, k
      $Err_j = O_j(1- O_j) \sum_k Err_k w_{jk}$
    End For
    For each weight w_{ij} in network
      Weight increment, $\Delta w_{ij} = (l) Err_j O_i$
      Weight update, $w_{ij} = w_{ij} + \Delta w_{ij}$
    End For
    For each bias Θ_j in network
      Bias increment, $\Delta \Theta_j = (l) Err_j$
      Bias update, $\Theta_j = \Theta_j + \Delta \Theta_j$
    End For
  End For
End While
End

Fig. 5.4 Back-propagation algorithm (Li et al. 2012)

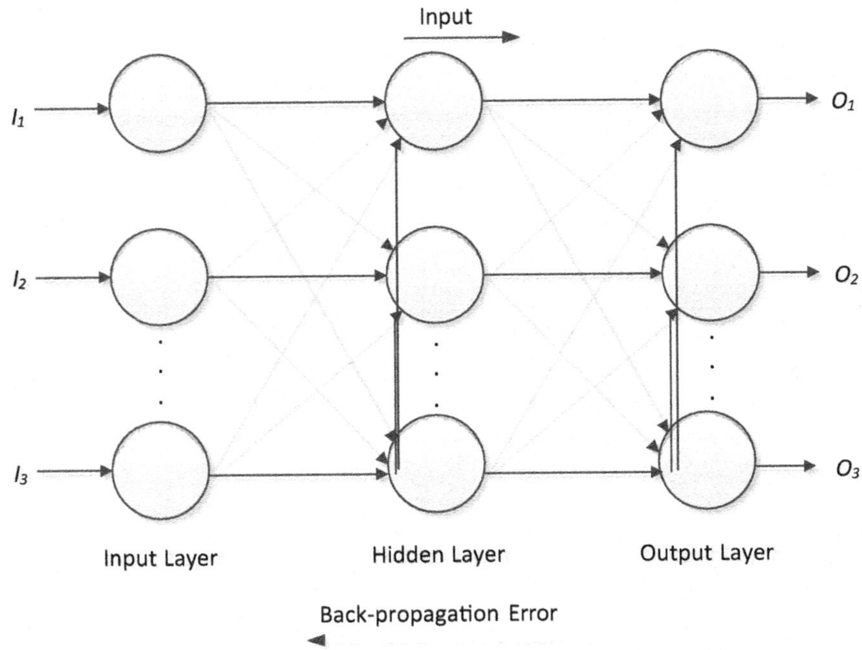

Fig. 5.5 Multi-layer back-propagation neural network structure

many other similar problems over the years. Many industries are trying to integrate DL nowadays due to its effectiveness in solving complex business problems.

As indicated earlier, meta-heuristic algorithm is an effective and efficient approach to optimize neural network as the algorithm can balance both exploration and exploitation. As a result, the complex and nonlinear problem can be solved through this approach. However, it is always better to improve, ensemble, or hybrid meta-heuristic to avail maximum outcome through meta-heuristic (Ojha et al. 2017). Although many meta-heuristic algorithms are used for training neural network, Biogeography-Based Optimization and Grey Wolf Optimizer can be some of the better choices for training MLP neural network (Mirjalili et al. 2014a, b).

5.3 Meta-heuristic Algorithms for Training Neural Network

As training or learning is an extremely vital process of neural network, it is quite challenging to determine the right algorithm for it. The training of neural network is an iterative process, and the prime objective of this learning is to minimize the loss function. Hence, the calculation is performed both forward and backward through each layer in the network. In light of training ANN, researchers attempted significantly toward the optimization of ANN parameters adopting various approaches like hybrid, ensemble, and so on. Gradient-based optimization method and back-propagation (BP) algorithm are other way of neural network learning that used to be implemented extensively as well. However, they suffer from limitations of getting trapped in local minima that create enormous opportunities for meta-heuristics algorithm to perform the learning efficiently. Neural Network can be trained by applying various meta-heuristic algorithms such as Genetic Algorithm, Particle Swarm Optimization, and Ant Colony Optimization. A number of meta-heuristic algorithms are growing enormously over the years. However, the current study concentrates mostly on Grey Wolf Optimizer for training neural network as per the scope of the research. Basically, learning algorithms need to be proposed with the aim of balancing two important characteristics called exploration and exploitation in addition to convergence ability and avoidance of local minima. Thus, algorithms should be experimented with whether these have the capability to produce global optimum.

ANN consists of numerous processing units, that is self-adapting and self-organizing. Consequently, it is able to perform real-time learning (Chong et al. 2021). The objective of the ANN learning is to ensure optimal network performance and various parameters of ANN such as number of input–output nodes, number of hidden layers, type of activation function, values of weight, and bias play an important role in this regard. Thus, various learning algorithms attempt to determine the optimal weight values as well as number of neurons. But, all the algorithms are not able to perform well due to various limitations that demand the investigation of meta-heuristic algorithms implementing the global optimization techniques to train the ANN so that the network can perform optimally. As meta-heuristic algorithms perform efficiently and show great potential for solving complex problems adopting various behaviors from nature, social, and physical laws, researchers continuously strive to optimize learning of ANN using these algorithms.

Grey Wolf Optimizer (GWO) is a kind of meta-heuristic algorithm initiated from grey wolves that mimics the leadership hierarchy and hunting mechanism of grey wolves in nature (Mirjalili et al. 2014b). The detail about this GWO algorithm will be explored extensively in later chapters of this book as this algorithm is applicable to challenging problems in unknown search spaces that have the capability to generate better result compared to other mate-heuristic algorithms like PSO, GSA, DE, EP, and ES (Faris et al. 2018; Gupta et al. 2015; Mirjalili et al. 2014b). Moreover, it can be applied efficiently for training multi-layer perceptron neural network and the empirical results demonstrate the true power of GWO, because the algorithm can faster decide the suitable thresholds, able to produce good classification rate, greater efficiency, and accuracy.

As GWO has emerged as an efficient meta-heuristic algorithm for solving complex optimization problem, the modification of this algorithm can be an effective strategy for solving the stock prediction problem and hence the emergence of the proposed algorithm entitled MGWO. Therefore, in the current research MGWO is an approach based on GWO for feature selection and prediction. Original GWO initializes the grey wolf population, updates the position of each search agent by the locations of the best solutions, and assesses the objective function of the algorithm. The MGWO algorithm is formed by improving the wolves attack strategy for the purpose of training ANN. The algorithm calculates the weights based on wolves fitness function and gives the highest weight to the dominant 4 wolf concurrently to improve the convergence, decide the suitable thresholds faster, and provide good classification rate, efficiency, and accuracy. For the modification of GWO, the necessary parameters are modified in comparison with original GWO and other algorithm, where the parameter value for maximum iteration and population size need to be maintained well for fair comparison.

5.4 Ensemble Model

In the context of the current work to optimize the performance of the neural network, the combination of multiple algorithms to improve the accuracy and stability of a classification model can be categorized as an ensemble model. Thus, the ensemble model consists a finite number of neural networks that can be trained to perform the same task. Research demonstrated that training the neural network through ensemble neural network model and combining the predictions can greatly improve the generalization capacity. Here, the neural network ensemble is a two-step process: firstly, by training a number of component neural networks and then, by combining the component predictions.

Researchers attempted for prediction applying ensemble algorithms acclaimed that an ensemble model built efficiently can concurrently make accurate prediction and manage the prediction errors well in diverse areas of the input space (Jothimani et al. 2016). The ensemble and similar approaches in financial prediction have been very popular and successful over the years (Sujatha and Punithavathani 2018; Niu et al. 2016; Lahmiri and Boukadoum 2015). Co-operative ensemble approach is the attention in this research where the prediction task can be divided into numerous sub-tasks to gain the prediction accuracy and the outcome of the prediction is sum of all sub-tasks. Figure 5.6 represents the architecture of an Ensemble Model, where input is forwarded to the sub-task 1, sub-task 2, and sub-task 3. The output produced by the various sub-tasks will be combined together and then forwarded to the final ensemble output.

As indicated in Fig. 5.7, ensemble model for ANN can be built to produce a better learning by strengthening generalization ability of ANN such that many neural

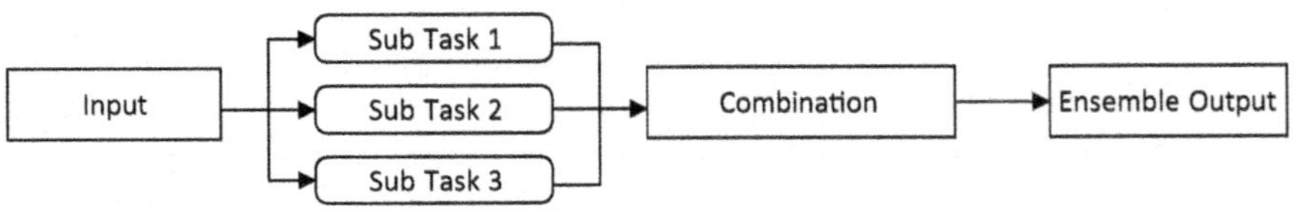

Fig. 5.6 Ensemble model architecture

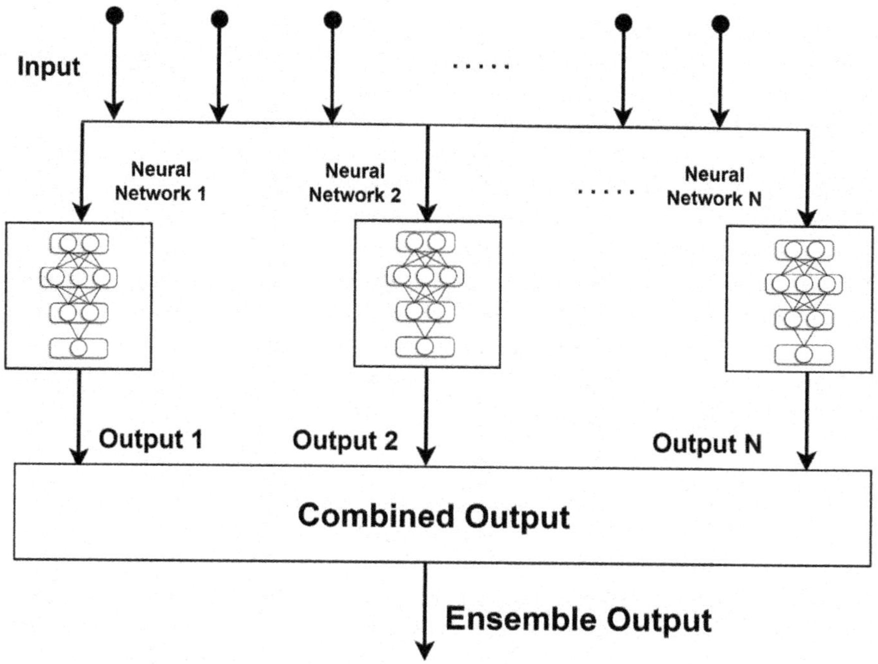

Fig. 5.7 An ensemble learning model for ANN

networks can be trained and predictions can be combined. Usually, ANN-based ensemble model can be produced through combination of numerous component ANN training and component predictions. Thus, a number of ANNs outputs are combined together to build the ensemble model, where the weighted average of each network output is the output of an ensemble model.

Evaluation of ensemble model performance is another important aspect that needs to be considered well while designing ensemble model. Many researches have proposed the evaluation techniques of ensemble model to their investigations (Hashino et al. 2007; Zhang et al. 2016; Das and Sengur 2010; Hosni et al. 2019). Evaluation of ensemble model needs to consider various measures such as predictive performance, computational complexity, comprehensibility, interpretability, scalability, usability, and robustness. Predictive performance can be measured using performance metrics, where Accuracy is the most frequently used as it is very simple. However, accuracy may not be always sufficient for ensemble model evaluation, especially when the dataset is imbalanced. For such scenario it is better to apply other ensemble model evaluation technique such as Recall, Precision, Specificity, and F-Measure. Computational complexity of ensemble model is another important criterion to be considered as the amount of lower CPU time is always expected. Usually, two complexity metrics are to be implemented in this regard to determine the training computation cost and the testing computation cost. Besides, compactness metric is another way of evaluating ensemble model that can be used for the evaluation of interoperability of an ensemble model, where number of classifiers and complexity of each of those classifiers can be used. Scalability of ensemble model can be ensured by building a model based on large amounts of data and usability of ensemble model can be evaluated based on a metric that can assess user's preference. Thus, an ensemble model should be designed in such a way that can have easily adjustable control parameters to ensure its efficiency.

The great benefit of ensemble model is that it can often produce better output than the basic or other integrated network. Consequently, the number of research is growing in this area as the training performance is getting better than the decorrelated individual network. Hence, the development of ensemble model consisting Modified Grey Wolf Optimizer (MGWO) and multi-layer perceptron (MLP)-based Artificial Neural Network is an extension of the previous researches that can be applied for the prediction of stock market.

5.5 Summary

This chapter has presented the general concept of neural networks and their training behavior. In addition, the use of meta-heuristic algorithm in training neural network was discussed in a section along with the generic concept of an ensemble model. The next chapter will intrude the readers with the overview concept of GWO algorithm and its use in training neural networks.

References

Balabanov T, Zankinski I, Dobrinkova N (2011, June). Time series prediction by artificial neural networks and differential evolution in distributed environment. In International Conference on Large-Scale Scientific Computing (pp. 198–205). Berlin, Heidelberg: Springer Berlin Heidelberg

Berry MJ, Linoff G (1997) Data mining techniques: for marketing, sales, and customer support. Wiley

Chong HY, Yap HJ, Tan SC, Yap KS, Wong SY (2021) Advances of metaheuristic algorithms in training neural networks for industrial applications. Soft Comput 25(16):11209–11233

Das R, Sengur A (2010) Evaluation of ensemble methods for diagnosing of valvular heart disease. Expert Syst Appl 37(7):5110–5115

Faris H, Aljarah I, Al-Betar MA, Mirjalili S (2018) Grey wolf optimizer: a review of recent variants and applications. Neural Comput Appl 30(2):413–435

Gupta P, Rana KPS, Kumar V, Mishra P, Kumar J, Nair SS (2015) Development of a grey wolf optimizer toolkit in LabVIEW™. In: 2015 international conference on futuristic trends on computational analysis and knowledge management (ABLAZE). IEEE, pp 107–113

Hashino T, Bradley AA, Schwartz SS (2007) Evaluation of bias-correction methods for ensemble streamflow volume forecasts. Hydrol Earth Syst Sci 11(2):939–950

Hosni M, Abnane I, Idri A, de Gea JMC, Alemán JLF (2019) Reviewing ensemble classification methods in breast cancer. Comput Methods Programs Biomed 177:89–112

Jothimani D, Shankar R, Yadav SS (2016) A comparative study of ensemble-based forecasting models for stock index prediction. In: Proceedings of the eleventh Midwest association for information systems conference (MWAIS), Milwaukee, Wisconsin, 19–20 May 2016, pp 1–7

Lahmiri S, Boukadoum M (2015) An ensemble system based on hybrid EGARCH-ANN with different distributional assumptions to predict S&P 500 intraday volatility. Fluct Noise Lett 14(01):1–10

Li J, Cheng JH, Shi JY, Huang F (2012) Brief introduction of back propagation (BP) neural network algorithm and its improvement. In: Advances in computer science and information engineering, 2:553–558. Springer Berlin Heidelberg

Mirjalili S, Mirjalili SM, Lewis A (2014a) Let a biogeography-based optimizer train your multi-layer perceptron. Inf Sci 269:188–209

Mirjalili S, Mirjalili SM, Lewis A (2014b) Grey wolf optimizer. Adv Eng Softw 69:46–61

Niu M, Wang Y, Sun S, Li Y (2016) A novel hybrid decomposition-and-ensemble model based on CEEMD and GWO for short-term PM2.5 concentration forecasting. Atmos Environ 134:168–180

Ojha VK, Abraham A, Snášel V (2017) Metaheuristic design of feedforward neural networks: a review of two decades of research. Eng Appl Artif Intell 60:97–116

Schmidhuber J (2015) Deep learning in neural networks: an overview. Neural Netw 61:85–117

Sujatha K, Punithavathani DS (2018) Optimized ensemble decision-based multi-focus imagefusion using binary genetic grey-wolf optimizer in camera sensor networks. Multimed Tools Appl 77(2):1735–1759

Wanto A, Zarlis M, Hartama D (2017) Analysis of artificial neural network backpropagation using conjugate gradient fletcher reeves in the predicting process. J Phys Conf Ser 930(1):012018

Wu J, Coggeshall S (2012) Foundations of predictive analytics. CRC Press, USA

Zhang Y, Phillips P, Wang S, Ji G, Yang J, Wu J (2016) Fruit classification by biogeography-based optimization and feed-forward neural network. Expert Syst 33(3):239–253

Chapter 6
Grey Wolf Optimizer: Foundations and Mathematical Models

Abstract This chapter introduces the Grey Wolf Optimizer (GWO), a nature-inspired meta-heuristic algorithm that emulates the leadership hierarchy and hunting behaviour of grey wolves in the wild. The algorithm balances exploration and exploitation by modelling alpha, beta, delta, and omega wolves as agents in the search space, with the top three guiding the optimization process. The chapter outlines the social structure and mathematical modelling of GWO, supported by equations and pseudocode. It also presents various modifications to enhance GWO's performance and addresses its limitations in complex, multimodal environments. Applications of GWO in feature selection and neural network training, particularly for classification tasks, are reviewed, showing improved convergence and accuracy compared to traditional algorithms. The chapter concludes by highlighting the value of hybrid or ensemble methods in overcoming GWO's tendency to get trapped in local minima, setting the stage for further advancements in optimized neural network training.

Keywords Grey Wolf Optimizer Algorithm · Alpha · Beta · Delta · Omega · Mathematical model of GWO · Variations of GWO · Ensemble of GWO

6.1 Introduction

Grey Wolf Optimizer (GWO) is a new meta-heuristic algorithm, which is emulated from grey wolves (*Canis lupus*), and it mimics the leadership hierarchy and hunting mechanism of grey wolves in nature (Mirjalili et al. 2014). The leadership hierarchy is simulated by applying four types of grey wolves, i.e. α, β, δ, ω, and hunting is implemented by applying three main steps, i.e. searching for prey, encircling prey, attacking prey. The GWO algorithm is applicable to challenging problems in unknown search spaces which produces better results than PSO, GSA, DE, EP, and ES (Faris et al. 2018; Gupta et al. 2015; Mirjalili et al. 2014).

GWO algorithm can be applied as a training algorithm for multi-layer perceptron. The algorithm improves the wolves attack strategy. It calculates the weights based on wolves fitness function and gives the highest weight to the dominant wolf

D. Das et al., *Optimization Algorithms in Machine Learning*, Engineering Optimization: Methods and Applications, https://doi.org/10.1007/978-981-96-3849-9_6

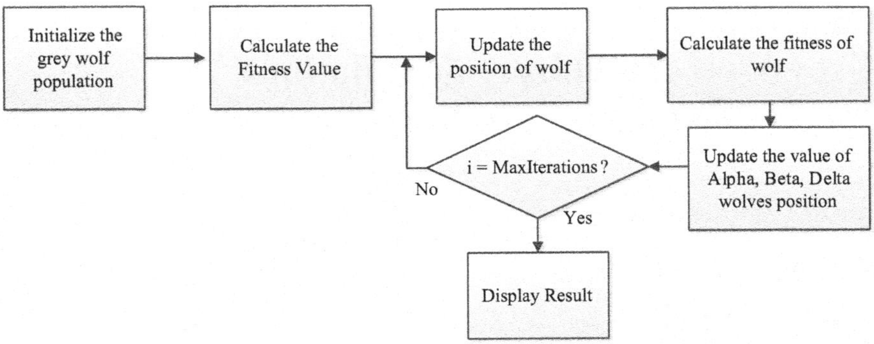

Fig. 6.1 Flow of GWO algorithm

concurrently to improve the convergence. The empirical results confirm the power of GWO and demonstrate that the algorithm can faster decide the suitable thresholds and provide good classification rate, efficiency, and accuracy. The flow of GWO algorithm is indicated in Fig. 6.1.

The solution of a problem through meta-heuristic algorithm needs to address two conflicting processes known as exploration and exploitation (Emary et al. 2018). The exploration process facilitates the algorithm to discover new areas into the problem search space through engaging abrupt alterations to the solutions. The promising areas may be explored to the search landscape and solution may be exempted from stagnation into local optimum through exploration. On the other hand, the exploitation process entails the algorithm to discover the neighboring zone so that expected solutions attained through exploration can be improved. Exploitation performs gradual adjustment to the solution so that the solution converges to the global optimum. GWO algorithm can be adjusted to make a good balance between both exploration and exploitation (Faris et al 2018; Mirjalili et al. 2014).

GWO is inspired through the searching by grey wolves in nature for optimal way of hunting preys and the algorithm is developed by Mirjalili et al. in 2014 (Mirjalili et al. 2014). The detail overview of GWO algorithm is explained in this section.

6.2 Basic Characteristics of Grey Wolf Optimizer Algorithm

Grey wolves (*Canis lupus*) belong to Canidae family, and they are considered as apex predators, meaning that they are at the top of the food chain. They have a very strict social dominant hierarchy. The algorithm divides the wolves into four types: α, β, δ, and ω, whereas each type of wolves displays the following social behavior.

The leaders are a male and a female, called alpha. Decision making about hunting, sleeping place, and time to wake, is made by alpha. Due to these reasons, alpha becomes the leader in the pack and others follow its orders. The best solution of a problem can be determined by identifying the location of alpha, as it is the best member in managing the pack.

The second level in the hierarchy of grey wolves is beta. The betas are subordinate wolves that help the alpha in decision making or other pack activities. The beta maintains the discipline of the pack that enforces the alpha's commands in the pack, and hence, it takes the role of the advisor to the alpha.

The third level in the hierarchy of grey wolves is delta and they need to submit to alphas and betas, but they dominate the omega. They are responsible for watching the boundaries of the territory, warning the pack in case of any danger, protecting and guaranteeing the safety of the pack, helping the alphas and betas when hunting prey, and providing food for the pack and caring for the weak, ill, and wounded wolves in the pack.

The last grey wolf in the hierarchy is omega and it may not be an important individual in the pack, but it has been observed that the whole pack faces internal fighting and problems in case of losing the omega, which is harmful to the group structure.

Group hunting is another interesting social behavior of grey wolves in addition to the social hierarchy. The main phases of grey wolf hunting are as follows: searching for the prey; tracking, chasing, and approaching the prey; pursuing, encircling, and harassing the prey until it stops moving; attacking toward the prey.

The fittest solution is considered as the alpha (α) in designing mathematical model of the social hierarchy of wolves for designing GWO (Mirjalili et al. 2014). Beta (β) and delta (δ) are respectively the second and third fittest solutions. Omega (ω) is considered for the rest of the candidate solutions. In the GWO algorithm, the hunting (optimization) is guided by α, β, and δ. The ω wolves follow these three wolves. GWO can be summarized in Fig. 6.2.

The original Grey Wolf Optimizer (GWO) algorithm is illustrated in Fig. 6.2, where, X_i represents the initial population of grey wolf; the GWO parameters such

Begin

Initialize the grey wolf population, X_i (i = 1,2,3,...,n)

Initialize a, A and C

Calculate the fitness of each search agent, where, X_α, X_β, X_δ are the best, second best and third best search agent consecutively

While i<*MaxIterations*

Update the position of current search agent for each search agent by equation

$$\vec{X}(t+1) = (\vec{X}_1 + \vec{X}_2 + \vec{X}_3)/3$$

Update a, A and C

Calculate the fitness of all search agents

Update X_α, X_β, and X_δ

$t = t + 1$

End While

Return X_α

End

Fig. 6.2 Pseudocode of GWO classification algorithm

a, A, C are the vectors; t represents the maximum number of iteration.

$$\vec{A} = 2\vec{a} \cdot \vec{r}_1 - \vec{a} \tag{6.1}$$

$$\vec{C} = 2 \cdot \vec{r}_2 \tag{6.2}$$

In Eq. (6.1), the values of $\vec{a}$ are linearly decreased from 2 to 0 over the course of iterations. At this stage, the estimation of the fitness for each search agents is made and the hunt agents are identified such as best hunt agent X_α, the second best hunt agent X_β, and the third best hunt agent $X_\eth$.

The updating of the location for the current hunt agent is made using the equation, $\vec{X}(i+1) = \left(\vec{X}_1 + \vec{X}_2 + \vec{X}_3\right)/3$. Then, the vectors are updated. Next, the fitness value of all hunts is estimated and the value for X_α, X_β, and $X_\eth$ are updated. The stopping condition is checked here to determine whether the iteration (t) reaches max number of iterations, if yes, then return and print the best value of solution X_α, otherwise, the algorithm will start through the same equation, $\vec{X}(i+1) = \left(\vec{X}_1 + \vec{X}_2 + \vec{X}_3\right)/3$.

6.3 Mathematical Model of Grey Wolf Optimizer Algorithm

GWO can be formed as per the mathematical equations below (Mirjalili et al. 2014):

$$\vec{D} = \left|\vec{C} \cdot \vec{X}_p(t) - \vec{X}(t)\right| \tag{6.3}$$

$$\vec{X}(t+1) = \vec{X}_p(t) - \vec{A} \cdot \vec{D} \tag{6.4}$$

where t denotes the current iteration, $\vec{A}$ and $\vec{C}$ are coefficient vectors, $\vec{X}_p$ is the position vector of the prey, and $\vec{X}$ denotes the position vector of a grey wolf.

The vectors $\vec{A}$ and $\vec{C}$ are calculated as follows:

$$\vec{A} = 2\vec{a} \cdot \vec{r}_1 - \vec{a} \tag{6.5}$$

$$\vec{C} = 2 \cdot \vec{r}_2 \tag{6.6}$$

where components of $\vec{a}$ are linearly reduced from 2 to 0 over the number of iterations and used for controlling the trade-off between exploitation and exploration. The following equations will be employed for updating the value of variable:

$$\vec{a} = 2 - t(2/X_i) \tag{6.7}$$

$$\vec{X}(t+1) = \left(\vec{X}_1 + \vec{X}_2 + \vec{X}_3\right)/3 \tag{6.8}$$

where X_i denotes the number of iterations, $\vec{r}_1$ and $\vec{r}_2$ are random vectors between [0, 1] which are employed to find the optimal solution. Appropriate idea about the potential location of prey can be availed by Alpha, Beta, and Delta, where they help the Omega to follow the suitable positions. The values of $\vec{X}_1$, $\vec{X}_2$, and $\vec{X}_3$ can be obtained through the equations below:

$$\vec{X}_1 = \vec{X}_\alpha - \vec{A}_1 \cdot \vec{D}_\alpha \tag{6.9}$$

$$\vec{X}_2 = \vec{X}_\beta - \vec{A}_2 \cdot \vec{D}_\beta \tag{6.10}$$

$$\vec{X}_3 = \vec{X}_\delta - \vec{A}_3 \cdot \vec{D}_\delta \tag{6.11}$$

In iteration t, the best 3 solutions are respectively, $\vec{X}_1$, $\vec{X}_2$, and $\vec{X}_3$. Where, the values of $\vec{D}_\alpha$, $\vec{D}_\beta$, and $\vec{D}_\delta$ are as below:

$$\vec{D}_\alpha = \left|\vec{C}_1 \cdot \vec{X}_\alpha - \vec{X}\right| \tag{6.12}$$

$$\vec{D}_\beta = \left|\vec{C}_2 \cdot \vec{X}_\beta - \vec{X}\right| \tag{6.13}$$

$$\vec{D}_\delta = \left|\vec{C}_3 \cdot \vec{X}_\delta - \vec{X}\right| \tag{6.14}$$

Exploration and exploitation in GWO can be expressed as follows:

Parameter $\vec{C}$ is the key element to facilitate exploration in terms of local optima stagnation as it contains random values between [0, 2] that offers random weights for prey to stochastically emphasize $C > 1$ and deemphasize $C < 1$. As a result, the solution inclines closer to the prey. However, parameter $\vec{A}$ is another source of exploration as the value of the parameter is controlled by a, that can be linearly declined from 2 to 0. The range of parameter $\vec{A}$ alters between the interval of [− 2, 2] as it contains random element. The value of $\vec{A} > 1$ and $\vec{A} < -1$ ensures exploration so that GWO algorithm starts searching globally. Conversely, the value of $\vec{A} > -1$ and $\vec{A} < 1$ ensures exploitation.

6.4 Variations of Grey Wolf Optimizer Algorithm

In the course of the most recent couple of years, different variations of GWO have been acquainted owing with various improvement issues. The modification of GWO algorithm has been proposed to comply with the difficult real-world optimization problem.

Due to the constraint of GWO to handle real-world problems, some modifications are proposed through update mechanism, some proposed to improve GWO operations, some proposed to enable the exploration and exploitation through ensemble or hybridization and some proposed to handle parallel computing platforms. This section is planned to give a quick overview about proposed GWO's variations and upgrades.

Mittal et al. (2016) proposed the improvement of GWO exploration through application of exponential decay function as indicated in Eq. (6.15). The approach recommended reducing the value of parameter a exponentially replacing linear modification. The proposed solution was tested over 27 benchmark functions and attained better result compared to other prominent meta-heuristic algorithms such as PSO, BA, CS, and GWO. Whereas, Long et al. (2017) investigated the ensemble of Modified Augmented Lagrangian (MAL) with Improved Grey Wolf Optimizer (IGWO) to adapt the parameter a applying the equation indicated in (6.16). The study attained a better result through the nonlinear adaptation with an appropriate balance between exploration and exploitation.

$$a = 2\left(1 - \frac{\text{Iteration}^2}{\text{MaxIteration}^2}\right) \tag{6.15}$$

$$a = \left(1 - \frac{\text{Iteration}}{\text{MaxIteration}}\right) \cdot \left(1 - \mu \cdot \frac{\text{Iteration}}{\text{MaxIteration}}\right)^{-1} \tag{6.16}$$

where μ is a nonlinear modulation index at the interval (0, 3).

In the recent time, GWO algorithm has been implemented for feature selection with the objective of selecting most appropriate features, decreasing number of features, and removing irrelevant, noisy, and redundant features. Li et al. (2017) investigated the ensemble of binary GWO and wrapper-based methods for feature selection. The study attempted to address medical diagnosis problem through application of a classifier called Kernel Extreme Learning Machine. Emary et al. (2016) proposed an ensemble of binary GWO and k-nearest neighbor (KNN) where GWO is applied as a feature selection approach. The study attained better result with faster convergence compared to GA and PSO. Another study for feature selection applying GWO was attempted by Emary et al. (2016). The study was successful to produce encouraging results with an option to avoid local minima.

Currently, another application of GWO attracted the researchers' attention in training neural network or ANN integrating GWO. The most common neural network is MLP, which is applied for classification. Mosavi et al. (2016) applied GWO-based training in combination with MLP for three different datasets and attained reasonable result compared to PSO, Gravitational Search Algorithm (GSA), and PSOGSA. Similar model was applied by Mohamed et al. (2015) for training MLP. The study was successful in producing lower error rate with faster convergence for MLP. In another study, Mirjalili et al. (2014) investigated the application of GWO to train MLP and produced better result in comparison with PSO, GA, ACO, ES, and PBIL.

As featured earlier, GWO has been utilized by numerous researchers because of its benefits over others, GWO depends on parameters which can balance between exploration and exploitation. Moreover, GWO is simple and flexible which utilizes basic analogy including the grey wolves in nature for hunting preys. Consequently, their usage is clear. At this juncture, the standard GWO has been demonstrated its potential for taking care of unimodal optimization issues, although when confronting complex multimodal optimization issues with substantial amount of local minima, the GWO is generally getting stuck into a local minima because of shortcoming of its population's decent variety (Faris et al. 2018).

Keeping in mind the shortcoming of GWO, the ensemble can be an effective option to enhance GWO's performance (Faris et al. 2018; Mosavi et al. 2016; Mohamed et al 2015). Notwithstanding the reality, ensemble can enhance the GWO's performance, an excess of ensemble may invoke more complicacy to the algorithm. Moreover, exploration and exploitation are the key operations to improve meta-heuristic algorithm's proficiency. In the literature, different types of approaches have been adopted by various algorithms such as crossover, mutation, and elitism operators in GA, random walks in CS, and parameter adaptation in GWO. To balance the exploration and exploitation, all meta-heuristic algorithm utilizes directly/indirectly a mechanism or operator.

6.5 Summary

The main objective of this chapter was to provide the readers with a comprehensive overview on the functionality of GWO algorithm so that the following chapters will be easy to be followed and understand the improvements were applied to achieve MGWO. Moreover, the recent implementation of GWO in feature selection and optimization is reviewed and discussed. Related meta-heuristic algorithm that has been used for the same purpose is discussed also. In the following chapter further review will be carried on to present the use of GWO in the literature for training neural networks for decision making and prediction.

References

Emary E, Zawbaa HM, Hassanien AE (2016) Binary grey wolf optimization approaches for feature selection. Neurocomputing 172:371–381

Emary E, Zawbaa HM, Grosan C (2018) Experienced gray wolf optimization through reinforcement learning and neural networks. IEEE Trans Neural Netw Learn Syst 29(3):681–694

Faris H, Aljarah I, Al-Betar MA, Mirjalili S (2018) Grey wolf optimizer: a review of recent variants and applications. Neural Comput Appl 30(2):413–435

Gupta P, Rana KPS, Kumar V, Mishra P, Kumar J, Nair SS (2015) Development of a grey wolf optimizer toolkit in LabVIEW™. In: 2015 international conference on futuristic trends on computational analysis and knowledge management (ABLAZE). IEEE, pp 107–113

Li Q, Chen H, Huang H, Zhao X, Cai Z, Tong C, Tian X (2017) An enhanced grey wolf optimization based feature selection wrapped kernel extreme learning machine for medical diagnosis. Comput Math Methods Med 2017:1–15

Long W, Liang X, Cai S, Jiao J, Zhang W (2017) A modified augmented Lagrangian with improved grey wolf optimization to constrained optimization problems. Neural Comput Appl 28(1):421–438

Mirjalili S, Mirjalili SM, Lewis A (2014) Grey wolf optimizer. Adv Eng Softw 69:46–61

Mittal N, Singh U, Sohi BS (2016) Modified grey wolf optimizer for global engineering optimization. Appl Comput Intell Soft Comput 8(1):1–16

Mohamed AAA, El-Gaafary AA, Mohamed YS, Hemeida AM (2015) Design static VAR compensator controller using artificial neural network optimized by modify grey wolf optimization. Paper presented on 2015 international joint conference on neural networks

Mosavi MR, Khishe M, Ghamgosar A (2016) Classification of sonar data set using neural network trained by gray wolf optimization. Neural Network World 26(4):393–415

Chapter 7
A Brief Review of Grey Wolf Optimizer: Variants and Applications

Abstract In this chapter, most exciting characteristics of Grey Wolf Optimizer (GWO) algorithm has been presented in terms of simplicity, flexibility, scalability, capability to yielding good convergence through the balance of exploration and exploitation, tuning of few parameters, and minimum information requirements for the initial search. The strength of GWO is revealed through analysis of various research articles in this chapter. Additionally, the chapter also summarizes how GWO algorithm can be applied for solving various optimization problems at present as well as in near future.

Keywords Grey Wolf Optimizer (GWO) · Variations of GWO · Updating mechanisms · Exploration · Exploitation · Hybridization of GWO · Ensemble of GWO · Parallelization of GWO · Multi-objective model

7.1 Introduction

Grey Wolf Optimizer (GWO) has been emerged as one of the latest Metaheuristics Swarm Intelligence (SI) algorithm that has been extensively applied for numerous optimization problems. Some of the most exciting characteristics of GWO are, but not limited to: simplicity, flexibility, scalability, capability to yield promising convergence through the balance between exploration and exploitation, few parameters to tune, and the initial search does not require any descent information. Consequently, the mentioned characteristics of GWO facilitates to place it as the center of research attraction within a short period.

Therefore, this review is an attempt to reveal the true power of GWO through the overview and summarization of various published researches. The review starts with the fundamentals of GWO in perspective of theoretical foundation and conceptual framework for optimization. The basic operations of GWO are also explained with the analysis of the variations of GWO through modification, hybridization, parallelization, and integration. Moreover, the implementation and improvements of GWO for the various fields such as machine learning, global optimization, engineering,

D. Das et al., *Optimization Algorithms in Machine Learning*, Engineering Optimization: Methods and Applications, https://doi.org/10.1007/978-981-96-3849-9_7

bioinformatics, networking, medical, image processing, open source software are discussed. Finally, the review concludes with summarization and future trends of GWO applications.

The Grey Wolf Optimizer (GWO) algorithm was proposed by Mirjalili et al. (2014a, b) in 2014, inspired by natural searching process to hunt the preys by grey wolves, can be applied for solving complex problems, because, the algorithm needs to control only two parameters to balance the exploration and exploitation that eventually escape it from being stagnant due to local optima. GWO algorithm contains four pack members namely, alpha, beta, delta and omega for performing the searching in an organized manner. The algorithm has been performed extremely well in comparison with benchmark problems and real-world case studies. Even though, the performance of the algorithm has not been investigated with the large or smaller sets of pack members that can deal with extensive real-world problems.

GWO is a unique population-based algorithm that is capable to produce better mathematical solution especially, the estimation of global optimum. Because, the algorithm can re-arrange the solution to different n-dimensional space that eventually mimics the natural hunting process of grey wolves. The algorithm consumes comparatively smaller amount of memory in comparison with Particle Swarm Optimization (PSO) where, the algorithm needs to function with position and velocity vectors.

Moreover, PSO limits only one best solution determined by all the particles, whereas, GWO determines three best solutions. In view of the simplicity, flexibility and other remarkable features of all SI based algorithms, GWO has taken its place as one of the best SI algorithm. Consequently, researchers are utilizing GWO algorithm more than other algorithms for solving optimization problems. Some of the best implementation of GWO algorithm for solving various complex optimization problems are, but not limited to, global optimization problems, electric and power engineering problems, scheduling problems, power dispatch problems, control engineering problems, robotics and path planning problems, environmental planning problems.

Applications of GWO can be categorized as three different ways:

(i) Application of GWO from Theoretical Perspective

Theoretically, GWO algorithm can be modified, hybridized, ensembled, parallelized, and tuned to multi-objective model. Figure 7.1 indicates the categorization of GWO from theoretical perspective.

(ii) Application of GWO for Solving Practical Problems

GWO is applied to solve various problems of engineering, medical, bioinformatics, machine learning, image processing, networking, and environmental over the years. Researchers are attempting to adapt the algorithm to the different other new areas relentlessly.

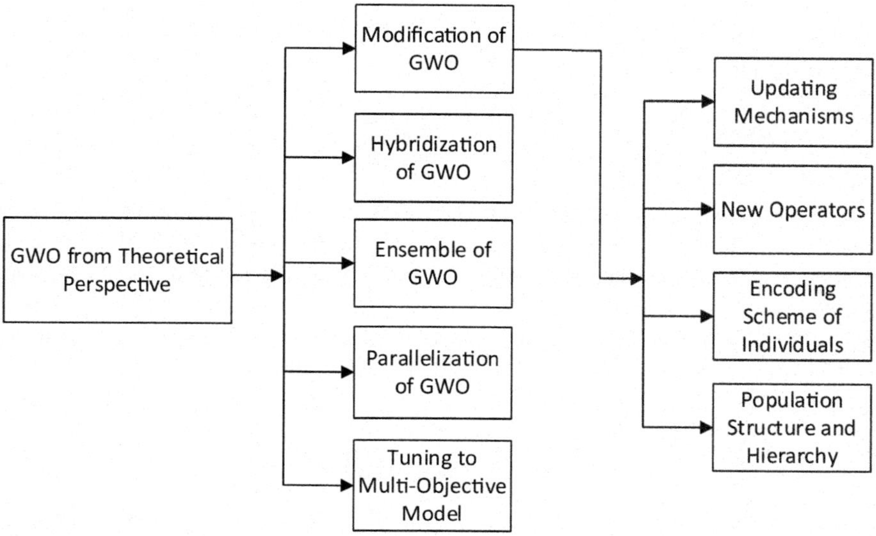

Fig. 7.1 Categorization of GWO from theoretical perspective

(iii) Extension of GWO for Creating Libraries

The application of GWO is not only limited to the theoretical and practical viewpoint, the algorithm can also be extended to support and produce libraries, toolbox, frameworks, open-source software.

7.2 Variations of GWO

GWO algorithm requires to be altered for the solution of complex problems, specifically, those need to be optimized. However, the alteration varies depending on the problems due to the nature of the issues to be handled for each problem such as updating the mechanism is implemented to handle the real-world problems with limitations, the improvement of GWO functioning is performed, the balance between the exploration and exploitation is performed through the hybrid version of GWO, and the modification of GWO is made to enhance the parallel computing.

The performance of GWO has been enhanced through the modification of:

(i) Updating Mechanisms
(ii) Inclusion of New Operators
(iii) Encoding Structure of the Individuals
(iv) Organization of Population and Hierarchy.

The main features of the mentioned alterations can be emphasized as follows:

Updating Mechanisms

This feature is conformed through the balance of exploration and exploitation where, the parameters of GWO is updated dynamically at one end and the updating of individuals are performed through various strategies at other end. Mittal et al. (2016) proposed the use of exponential decay function rather than linear one to decrease the value of a so that the exploration by GWO can be enhanced as indicated in Eq. (7.1) where, *Iteration* indicates the existing location and maximum number of iterations are denoted by *MaxIteration*. The proposed method achieved improved exploration where, the method is evaluated through the comparison of 27 benchmark functions and 3 different methods namely, CS, PSO, and BAT (Mittal et al. 2016).

$$a = 2\left(1 - \frac{\text{Iteration}^2}{\text{MaxIteration}^2}\right) \tag{7.1}$$

Similar approach has been adopted by Long and Xu (2016) that proposed to nonlinearly update the parameter a as indicated in Eq. (7.2) where, μ is a nonlinear modulation index at the interval (0, 3). Eventually, the nonlinear updating method improved the balance of both exploration and exploitation (Long and Xu 2016).

$$a = \left(1 - \frac{\text{Iteration}}{\text{MaxIteration}}\right) \cdot \left(1 - \mu \cdot \frac{\text{Iteration}}{\text{MaxIteration}}\right)^{-1} \tag{7.2}$$

On the other hand, Rodríguez et al. applied fuzzy logic to modify the a and C parameter through the dynamic adaptation (Rodríguez et al. 2016, 2017). The proposed approach has been adopted by Yogapriya and Nithya (2018) for reducing the high dimensional texture features and determining the best features in designing an effective medical image retrieval system (Yogapriya and Nithya 2018).

However, the alteration is different from the modification proposed by Kumar et al. (2017) and Long et al. (2018) where, a parameter was updated over the course of iteration and C parameter was updated dynamically over the number of iterations (Kumar and Kumar 2017; Long et al. 2018).

Another model is designed by Dudani and Chudasama (2016) that updates the wolves' position through the integration of step size proportional to the fitness of every individual as indicated in Eq. (7.3) where, the step size of ith dimension in tth iteration is, X_i^{t+1} and the fitness value is, $f(t)$. The advantage of the proposed approach over the other is, in terms of the requirement of number of parameters and defining initial parameter. Because, the model needs less parameters and moreover, initial parameter is not necessary to define. The designed model has been evaluated through the comparison of 21 benchmark functions and it produced comparatively faster convergence (Dudani and Chudasama 2016).

$$X_i^{t+1} = \left(\frac{1}{t}\right)^{|(\text{best} f(t) - fi(t))/(\text{best}(f(t) - \text{worst} f(t)))|} \tag{7.3}$$

An algorithm entitled wdGWO has been designed by Malik et al. (2015) that updates the location of each individual operators α, β, and δ through the calculation of weighted sum of best locations instead of simple average. In proposed algorithm, the best operator receives a weight through the multiplication of corresponding A and C. The proposed approach is evaluated with the set of benchmark functions and other SI algorithms where, the performance of the algorithm demonstrated significantly better than other counterpart (Malik et al. 2015).

Rodríguez et al. (2017) proposed a model where, the decision of adjusting the candidate parameters are performed dynamically and hence the actual parameters are selected based on the considerable effects on the functioning of the algorithm. Here, the position of ω operator is updated through the weighted average, fitness and fuzzy logic consecutively. The evaluation of the proposed model with the benchmark functions determined that the fuzzy logic-based model performs better than other similar models (Rodríguez et al. 2017).

Inclusion of New Operators

The performance of GWO is improved through the inclusion of new operators such as crossover and application of local search technique. In light of the similar approach, Kishor and Singh (2016) attempted to design a model that integrated a crossover operator for two selected individuals randomly so that information can be shared among the mates in every single pack. The model was evaluated with the original GWO and six benchmark functions. The evaluation eventually confirmed that the application of crossover operator enhances the ability to produce better solution and faster convergence (Kishor and Singh 2016). Chandra et al. (2016) applied this modified form of GWO to enhance the Web Services Quality and hence determined the optimal solution that is better than original GWO and GA (Chandra et al. 2016).

Removal of poor search agents from GWO and relocate the position of the wolves α, β, and δ so that exploitation can be enhanced. The design of such model is known as Evolutionary Population Dynamics (EPD) where, GWO randomly reinitializes the worst search agents so that exploration can be enhanced as well. The model had been proposed by Saremi et al. (2015) where, the evaluation of the model has been conducted with 13 different test functions and determined that the proposed model enhanced the exploration, exploitation, convergence, and avoidance of local optima for the GWO algorithm significantly (Saremi et al. 2015).

Other forms of GWO algorithm improvement through the inclusion of new operators include, Zhou et al. (2016) attempted to tune the parameters of GWO with chaotic local search for the small hydro generator cluster that demonstrated the good accuracy of result in engineering practice (Zhou et al. 2016). Rodríguez et al. (2017) attempted for the modification of GWO parameters through fuzzy hierarchical operator (Rodríguez et al. 2017). Mahdad and Srairi (2015) joined GWO algorithm with pattern search algorithm that ultimately applied for the security smart grid power system management problems in critical situations (Mahdad and Srairi 2015).

Zhang and Zhou (2015) designed an extension of GWO algorithm to solve clustering problems, known as Powell local optimization algorithm (PGWO). Consequently. PGWO algorithm produced significant result with the improvement of GWO algorithm and that was determined by evaluating the proposed algorithm with other similar evolutionary algorithms (Zhang and Zhou 2015).

Encoding Structure of the Individuals

An encoding model entitled Complex Valued-Encoding Grey Wolf Optimization Algorithm has been designed by Luo et al. (2016). The model proposed to apply two sets of genes namely: real and imaginary to express genes and hence extend the capacity of the individual. Here, the objective function is selected by the modules and the angles. Consequently, the individual and diverse population of GWO have been enhanced in this model so that local optima can be avoided. Finally, the enhanced GWO algorithm has been evaluated using 16 different functions and concluded that the modified algorithm with complex valued-encoding can be faster in convergence, accurate and robust that can be applied for solving practical problems efficiently (Luo et al. 2016).

Organization of Population and Hierarchy

The population and hierarchy of the GWO algorithm can be re-structured because the algorithm consists a distinct population structure that is hierarchical as well. Eventually, such structure of the GWO algorithm stimulated to further investigate on the modification of the population and hierarchy. As an attempt to support the above concept, Yang et al. (2017) proposed a model that separated GWO algorithm into two distinct subpopulations: cooperative hunting group, and random scout group respectively.

The cooperative hunting group ensures an efficient hunting by the four operators α, β, δ, and ω through its hierarchical collaboration and three different operations namely, searching for prey, encircling prey and attacking prey in an unfamiliar search space. However, random scout group carries out a global search randomly and maintains the balance between the exploration and exploitation to avoid local optima. The model is validated with three different case studies and determined that the organization of population and hierarchy brought improved global convergence, efficient power tracking and enriched fault ride to GWO algorithm compared to other contemporary heuristic algorithms (Yang et al. 2017).

Hybridization of GWO

In metaheuristics world, hybridization denotes to the linking of two or more algorithms to achieve the benefits and power of each algorithm in one hand and minimize the substantial weaknesses at other hand. In view of hybridization, many researchers attempted to hybridize GWO with other metaheuristic algorithms to improve the overall performance of GWO algorithm in literature.

Singh and Singh (2017) attempted to hybridize GWO with PSO algorithm to improve both the exploration and exploitation that eventually improves the convergence. The performance of the hybrid version of the algorithm has been evaluated with other similar unimodal, multimodal, and fixed-dimension multimodal functions and determined that hybrid version of GWO can significantly improve the quality, stability, convergence speed, and finding the global optimum for the solution (Singh and Singh 2017).

Other attempts to hybridization of GWO include, Zhu et al. (2015) designed a hybrid model consists of GWO and Differential Evolution (DE) that can update the best position of α, β, and δ. In this model, DE compels GWO to skip the stagnation through its robust searching capability (Zhu et al. 2015).

Singh and Hachimi (2018) modeled hybrid GWO using Mean Grey Wolf Optimizer (MGWO) and Whale Optimizer Algorithm (WOA) that can solve constrained nonlinear optimization function efficiently (Singh and Hachimi 2018). Al-Tashi et al. (2019) proposed a hybrid GWO and PSO algorithm to form a binary version that can improve the accuracy, chooses the best optimal features, and reduces the computational time for 18 different datasets collected from UCI repository (Al-Tashi et al. 2019).

Ensemble of GWO

In the context of the metaheuristics, ensemble is the combination of multiple algorithms to improve the accuracy and stability of an algorithm. In light of ensemble of GWO, researchers attempted to strengthen the performance of GWO algorithm through the ensemble of the algorithms with other algorithms. Moreover, the ensemble based approaches have been popular and successful in recent years (Sujatha and Punithavathani 2018; Niu et al. 2016; Lahmiri 2017).

Long et al. (2017) proposed an ensemble model consisting of Modified Augmented Lagrangian (MAL) with Improved Grey Wolf Optimizer (IGWO) to achieve a better result through the nonlinear adaptation with an appropriate balance between exploration and exploitation. Other forms of ensemble algorithms are, Li et al. (2017) designed the ensemble of binary GWO and wrapper-based method for feature selection that can address medical diagnosis problem.

Emary et al. (2016) modeled an ensemble of binary GWO and k-nearest neighbor (KNN) where GWO is applied as a feature selection approach and the model produced encouraging results that could also avoid local minima. Dai et al. (2018) designed an ensemble model entitled CEEMDAN-MGWO-SVM (Complete Ensemble Empirical Mode Decomposition with Adaptive Noise and Support Vector Machine Optimized by Modified Grey Wolf Optimization Algorithm) that can efficiently perform the daily peak load forecasting for power system. Moreover, the ensemble model demonstrated its potential for prediction that can ultimately effect the formulation of power generation plan, power grid dispatching, power grid operation and power supply reliability of power system (Dai et al. 2018).

Parallelization of GWO

The process of executing multiple computations simultaneously to make processing faster through multiple processing units is known as Parallelization. In metaheuristics, parallelization can be achieved through the separation of the population into numerous subpopulations so that each subpopulation can progress on a different processor. This will result the solution to be better in quality and decrease the execution time.

In view of the implementation of GWO algorithm for Parallelization, Pan et al. (2015) proposed a parallelized GWO algorithm that could be applied for the solution of numerical optimization problems. The algorithm divided the population wolves α, β, δ, and ω into numerous independent subpopulations, maintaining the original structure of the GWO. In this structure, the wolves are able to communicate to each other residing in different groups and the information is circulated accordingly. The proposed parallel version of GWO is then evaluated with four different benchmark functions and achieved better solution in terms of speed, accuracy and convergence (Pan et al. 2015).

Chen et al. (2019) initiated a parallel version of GWO algorithm for feature selection due to the limitations of: longer execution time for single-machine execution and computational complexity of higher dimensional data. The study proposed a Spark-based platform consisting of parallel binary GWO for feature selection so that processing of large volume of data becomes simpler, faster and storage requirement is minimized (Chen et al. 2019).

Tuning to Multi-objective Model

Multi-Objective model involves handling more than one objective function simultaneously. Most of the versions of GWO algorithm proposed in literature, involved in handling single-objective function only. However, some GWO based models attempted to solve multi-objective optimization problems as well. Mirjalili et al. (2016) brought a modification to GWO with integration of fixed-sized external archive that can outline the social hierarchy and select α, β, and δ wolves to design a multi-objective GWO.

The evaluation of the multi-objective model with two different multi-objective model established that GWO based multi-objective model performs better (Mirjalili et al. 2016). Hao and Tian (2019) applied multi-objective GWO algorithm for the prediction of wind power forecasting and the algorithm demonstrated good accuracy and stability (Hao and Tian 2019).

7.3 Applications of GWO for Solving Practical Problems

GWO has demonstrated the remarkable performances to handle the practical applications since the inception of the algorithm and hence, the review attempts to reveal the application of the algorithm into various practical applications explicitly, machine learning, engineering, wireless sensor network, environmental, medical, bioinformatics, and image processing applications.

GWO in Machine Learning

Machine-learning-based problems can be solved successfully applying GWO algorithm. Problems and related issues solvable by GWO algorithm can be explained as:

Feature Selection, Training Neural Networks, Optimization of Support Vector Machines, and Clustering Applications are some of the vital machine-learning issues well handled by GWO.

In data mining and machine learning, selection of appropriate features for classification and prediction is a very essential and challenging issue. Because, the performance of the prediction is greatly influenced by appropriate features. Hence, feature selection has to be dealt with selecting most appropriate features, reducing number of features, excluding noisy, irrelevant, and redundant features.

In selection of the most suitable features, Li et al. (2017) employed binary GWO and wrapper-based method for feature selection that eventually address the medical diagnosis problem through application of a classifier called Kernel Extreme Learning Machine. Another feature selection attempt was initiated by Emary et al. (2016) where, GWO algorithm is employed for feature selection along with ensemble of binary GWO and k-nearest neighbor (KNN). Evaluation of the proposed model with GA and PSO based approach determined that GWO can facilitate faster convergence, produce encouraging results, and avoid local minima.

Tu et al. (2019) modified GWO algorithm and proposed Multi-Strategy Ensemble GWO (MEGWO) algorithm that could effectively perform the feature selection for the real-world machine-learning-based optimization problems. The proposed algorithm brought three modifications to GWO algorithm namely, global-best leading strategy is enhanced to provide better local search, adaptable cooperative strategy is improve to support global search, and foraging strategy is disseminated to balance between exploration and exploitation (Tu et al. 2019).

Training the Artificial Neural Network (ANN) is another most vital machine-learning issue. Because, the performance of ANN depends on the weights and structure of the network and the appropriate structure is determined through training that eventually contributes to generate better prediction result. In this regard, Mirjalili (2015) attempted to apply GWO for training multi-layer perceptron of ANN that availed high exploration and exploitation with the improvement of local optima avoidance. In a different study, Nur and Ülker (2018) proposed the application of GWO for optimizing ANN and hence designed a hybrid cloud-based Intrusion Detection and Response System (IDRS). The study produced good result that can successfully detect intrusion over the cloud. Parsian et al. (2017) proposed an optimization of ANN using GWO for melanoma detection where, GWO trained ANN to determine the optimal initial weights and the outcome is really encouraging.

Turabieh (2016) designed an optimization of ANN through GWO to predict heart disease where, the study achieved better prediction result for the heart disease related medical dataset. Mosavi et al. (2016) proposed classification approach of sonar dataset applying GWO-ANN where, GWO is used for training ANN. Similar kind of ANN optimization through application of GWO was proposed by Mohamed et al. (2015) for designing the static VAR compensator (SVC) controller. The study produced better outcome with lower error values and faster convergence. Hence, GWO is an excellent algorithm that can be applied effectively for training ANN.

Support Vector Machine (SVM) is an efficient supervised learning algorithm for classification in machine learning. SVM parameters can be optimized through algorithms like GA, PSO, BAT, FA, and GWO. However, the accuracy of the classification depends heavily on the learning that needs proper investigation to determine an appropriate training algorithm (Faris et al. 2018; Wang et al. 2016; Mirjalili et al. 2014a, b). Eswaramoorthy et al. (2016) applied GWO algorithm for training SVM through the tuning of gramma and sigma parameters for classification of intracranial electroencephalogram signals where, the study produced better result in comparison with other classifier.

Mustaffa et al. (2015) attempted to optimize Least Squares Support Vector Machines (LSSVM) parameters through GWO algorithm for the commodity time series data where, the study produced better result with reduced error rate compared to LSSVM optimized by Artificial Bee Colony (ABC) algorithm (Mustaffa et al. 2015). Elhariri et al. (2016) proposed an EMG signals classification system that implemented GWO algorithm to produce better classification through SVM. The study gained better classification accuracy compared to the typical SVM using the radial basis function (RBF) kernel function (Elhariri et al. 2016).

Another machine learning and data mining issue called Clustering that separates the data into various groups known as clusters and places the data with similar features in same cluster. Clustering has been performed through the implementation of various metaheuristic algorithms over the years. GWO algorithm has brought a new avenue for clustering which can perform better than k-means clustering because the algorithm facilitates means to avoid local minima. In view of clustering in machine learning, Kumar et al. (2017) proposed GWO algorithm based clustering called GWAC where, GWO is applied to determine the optimal cluster centers in a search space. The evaluation of the proposed algorithm with six different algorithms indicates that, the performance of GWO based clustering is better than other algorithms that can also provide better support to avoid being trapped in local minima (Kumar et al. 2017). Zhang and Zhou (2015) designed an optimization model called PGWO based on Powell Local Optimization where, GWO algorithm is applied effectively for clustering (Zhang and Zhou 2015).

GWO is not only capable to solve the discussed machine-learning problems, it can also tackle all the recent machine-learning-based issues.

GWO in Engineering Applications

In the world of optimization, engineering is the most complex one, because the day-to-day life of human is directly related to the newer innovations through various engineering applications. Surprisingly, GWO algorithm has demonstrated great potential in handling the real-world engineering issues. Some of the engineering applications where GWO algorithm has been implemented effectively are as follows: power dispatching, controller designing and tuning, robotics and path planning, and also various other engineering purposes.

Power dispatching is a nonconvex and nonlinear optimization problem where, optimal load dispatch has to be determined for the optimal distribution of the available resources. Sulaiman et al. (2015) proposed GWO algorithm-based model to solve optimal reactive power dispatch (ORPD) issues where, GWO has been applied to determine the best combination of control variables namely, generator voltages, tap changing transformers' ratios, and the amount of reactive compensation devices. The proposed model is tested with two different case studies called IEEE 30-bus system, IEEE 118-bus system and then evaluation with other systems resolved that, GWO based model has demonstrated notable performances (Sulaiman et al. 2015).

An attempt for Economic Load Dispatch (ELD) implementing GWO algorithm to optimize the operating strategy has been initiated by Pradhan et al. (2016) where, GWO provided the optimal solution for nonlinear characteristics like ramp rate limits, valve point discontinuities and prohibited operating zones of generator. The implementation of proposed model with 10, 40, 80, and 140 units and evaluation of GWO based solution with similar models in literature settle that GWO provides better solution (Pradhan et al. 2016).

Designing and Tuning controllers is another optimization problem in engineering that is receiving the researchers' attention recently where, GWO algorithm has been implemented effectively for tuning Integral (I), Proportional-Integral (PI), and Proportional-Integral-Derivative (PID) parameters of controllers. For fine-tuning the PI parameter of controllers, Li and Wang (2015) attempted to implement GWO algorithm and observed better solution compared to Z-N Engineering Tuning, GA, and PSO. The proposed solution can be used for heat exchange system in nuclear, thermal, and marine power plant (Li and Wang 2015). Yadav et al. (2016) modeled an optimization of PID parameter through GWO algorithm that can regulate the ball position of the magnetic levitation system (MLS) for controllers. The validation of the proposed system with classical tuning indicates that GWO based PID optimization improves both time and frequency in addition to minimize the errors (Yadav et al. 2016).

Robotics and path planning is a complex optimization in engineering where the path for the robot has to be determined in such a way that it passes over all the points in a chosen area and the robots can be deployed to various applications like vacuum cleaning, painting, autonomous underwater vehicles creating images mosaics, demining, lawn mowers, automated harvesters, window cleaners, inspection of complex structures, and various other purposes. GWO algorithms can be applied effectively for solving the complexity of path planning for robots.

Kamalova et al. (2019) proposed the application of multi-objective version of GWO algorithm that can explore a new place and enhance the accuracy of map simultaneously in an unknown space that is known as multi-robot exploration. The proposed GWO based multi-objective optimization demonstrated good performance where, the adjustment between two objective function is maintained through Pareto-optimal solutions (Kamalova et al. 2019). Another study is undertaken by Dewangan et al. (2019) for the path planning of 3D multi-Unmanned Aerial Vehicle (UAV) that applied GWO algorithm for the optimization of a path between start point and end point. The evaluation of the proposed GWO based optimization with Dijkstra, BAT,

BBO, PSO, GSO, WOA, SCA ascertains that the performance of GWO is appropriate algorithm in path planning for 3D multi-UAV (Dewangan et al. 2019).

GWO in Wireless Sensor Network

In Wireless Sensor Network (WSN), the components are self-organized and most orderly distributed. Hence, there are various issues need to be solved for WSN such as location and tracking, deployment, coverage, quality of service (surveillance), and so on. GWO algorithm can be applied to deal with various issues related to WSN. Shieh et al. (2016) adopted GWO algorithm to solve the node localization problem of WSN where, the GWO algorithm determines the appropriate position of unknown nodes.

The evaluation of GWO algorithm in comparison with PSO and Modified Bat Algorithm (MBA) indicates that the performance of GWO algorithm is better than other algorithms for optimizing the computation time, percentage of localized node, and minimizing localization error (Shieh et al. 2016). Al-Aboody and Al-Raweshidy (2016) designed a three-level hybrid clustering protocol (MLHP) for WSN applying GWO algorithm where, the implementation of GWO algorithm ensures the optimization of routing for data transfer. The validation of GWO based model demonstrated better performance compared to other algorithms in terms of longer network lifetime, stability period, and more residual energy (Al-Aboody and Al-Raweshidy 2016).

GWO in Medical and Bioinformatics Application

In medical and bioinformatics, facilitating the health professionals applying the latest optimization technique is the most fascinating research area. GWO algorithm can be applied effectively for the various issues related to medical and bioinformatics. Earlier Diagnosis of disease is the most vital for healthcare to apply appropriate treatment and GWO has demonstrated potential at this area.

El Bakrawy (2017) proposed a Heart Disease Diagnosis algorithm applying GWO and Naïve Bayes Classifier (NB) where, GWO has been implemented for the selection of attributes weights for NB that can eventually optimize the classification accuracy. The proposed algorithm achieved better result for the UCI Machine Learning Cleveland Heart Database compared to single NB classifier algorithm (El Bakrawy 2017).

Another Diagnosis model is proposed by Sharma et al. (2019) where modification of GWO algorithm is performed to design a model that can diagnose the Parkinson's disease at earlier stage. The role of GWO algorithm to the proposed system includes searching for appropriate features and select accordingly for the prediction that performs better than other feature selection algorithms like Random forest, k-nearest neighbor classifier and decision tree.

Overall, the GWO based prediction algorithm performed better than the existing Parkinson's disease prediction system available in literature (Sharma et al. 2019). Hence, it is evident that GWO algorithm can be applied to the different areas of health and bioinformatics through appropriate investigation.

GWO in Image Processing

Image processing is another optimization technique where operations are needed to perform on an image so that an enhanced image can be produced or valuable features can be extracted for the image. Hence, image processing is a kind of signal processing where, image is a type of input and the output of the image can be another image or features related to that image.

GWO algorithm has been also demonstrated well performance in the field of image processing with appropriate optimization. Pathak et al. (2019) applied GWO algorithm for an Image Steganalysis model which is a technique to determine the available hidden messages from the cover images. The proposed model produced better result because of substantial contribution by GWO algorithm in determining the appropriate features for steganalysis from set of original features (Pathak et al. 2019).

An image processing application that implemented GWO algorithm to produce high-resolution image from low-resolution image and the model has been proposed by Rajput et al. (2019). The proposed model applied GWO algorithm that has demonstrated notable performances for minimizing least square representation (LSR), an optimization technique to obtain appropriate reconstruction weights (Rajput et al. 2019). Many other image processing applications have been undertaken over the years that implemented GWO algorithm.

7.4 GWO for Open Source Software

In recent times, various types of frameworks and libraries are created based on nature-inspired optimization. Those frameworks and libraries can be used by the rapidly growing scientific community to create a robust set of optimizers as open source software. The implementing of GWO algorithm will increase the popularity and portability among researchers. Moreover, the powerful libraries and packages through the application of metaheuristic algorithm like GWO will enhance the capability of solving complex problems on a much higher scale. The inventor of this GWO an open Matlab version of GWO immediately after the publication of the article consisting original GWO algorithm.

Gupta et al. (2015) created a Toolkit implementing GWO algorithm for LabView that is an excellent environment for industry practitioners for measurement and control. The inclusion of GWO enriched the optimization of LabView environment because, Differential Evolution (DE) was the only available optimization technique earlier (Gupta et al. 2015). EvoloPy-FS Framework is a Python open-source optimization platform created by Khurma et al. (2020) that can be used for optimized feature selection in critical data mining and machine learning. The framework includes various SI algorithms including GWO for facilitating researchers to resolve various machine-learning issues and visualize the results with minimum programming efforts (Khurma et al. 2020).

7.5 Summary

Taking into consideration the above discussion, wide variety of problems have been well optimized by GWO algorithm compared to many other SI algorithm since the invention of the algorithm. The main motivation of the application of GWO algorithm for various optimization problem consists of the simplicity, easy tuning of the parameter, and flexible exploratory nature. However, GWO contains certain limitations and drawbacks, similar to other SI algorithms.

As indicated in NFL theorem, no single algorithm exists in literature that can solve all optimization problems. Hence, GWO algorithm cannot be applied for solving all optimization problems. But, the algorithm can be altered to adapt with many real-world optimization problems.

Some of the limitations of the GWO algorithms include: the algorithm is suitable for single-objective problems although it can be enhanced for multi-objective problems; Low capacity to solve the issues of multimodal search landscape because, three operators α, β, and δ converge to the same solution; addition of extra random components for the mutation of solutions may determine global optimum; more number of variables may degrade the performance of GWO; faster convergence and exploitation may produce local solutions; if the algorithm is trapped to local minima, there is no automatic mechanism to decelerate the convergence; poor search agents among α, β, and δ wolves of GWO deteriorate exploitation; poor search agents around search space worsen exploration; the positions of wolves are updated mostly based on the experience of alpha, beta, and delta leaders in GWO that may lead to premature convergence.

However, most of the limitations of GWO algorithm can be opted out through appropriate tuning of the algorithm. Finally, modification of the GWO algorithm is the best strategy based on the problems encountered and eventually, the algorithm will be promising to solve many real-world applications.

References

Al-Aboody NA, Al-Raweshidy HS (2016).Grey wolf optimization-based energy-efficient routing protocol for heterogeneous wireless sensor networks. In: 2016 4th international symposium on computational and business intelligence (ISCBI), Sept 2016. IEEE, pp 101–107

Al-Tashi Q, Kadir SJA, Rais HM, Mirjalili S, Alhussian H (2019) Binary optimization using hybrid grey wolf optimization for feature selection. IEEE Access 7:39496–39508

Chandra M, Agrawal A, Kishor A, Niyogi R (2016) Web service selection with global constraints using modified gray wolf optimizer. In: 2016 international conference on advances in computing, communications and informatics (ICACCI), Sept 2016. IEEE, pp 1989–1994

Chen H, Han L, Hu Z, Hou Q, Ye Z, Zeng J, Yuan J (2019) A feature selection method of parallel grey wolf optimization algorithm based on Spark. In: 2019 10th IEEE international conference on intelligent data acquisition and advanced computing systems: technology and applications (IDAACS), Sept 2019, vol 1. IEEE, pp 81–85

Dai S, Niu D, Li Y (2018) Daily peak load forecasting based on complete ensemble empirical mode decomposition with adaptive noise and support vector machine optimized by modified grey wolf optimization algorithm. Energies 11(1):163

Dewangan RK, Shukla A, Godfrey WW (2019) Three dimensional path planning using grey wolf optimizer for UAVs. Appl Intell 49(6):2201–2217

Dudani AR, Chudasama K (2016) Partial discharge detection in transformer using adaptive grey wolf optimizer based acoustic emission technique. Cogent Eng 3(1):1256083

El Bakrawy LM (2017) Grey wolf optimization and naive Bayes classifier incorporation for heart disease diagnosis. Aust J Basic Appl Sci 11(7):64–70

Elhariri E, El-Bendary N, Hassanien AE (2016) A hybrid classification model for EMG signals using grey wolf optimizer and SVMs. In: The 1st international conference on advanced intelligent system and informatics (AISI2015), Beni Suef, Egypt, 28–30 Nov 2015. Springer, Cham, pp 297–307

Emary E, Zawbaa HM, Hassanien AE (2016) Binary grey wolf optimization approaches for feature selection. Neurocomputing 172:371–381

Eswaramoorthy S, Sivakumaran N, Sekaran S (2016) Grey wolf optimization based parameter selection for support vector machines. COMPEL Int J Comput Math Electr Electron Eng 35(5):1513–1523

Faris H, Aljarah I, Al-Betar MA, Mirjalili S (2018) Grey wolf optimizer: a review of recent variants and applications. Neural Comput Appl 30(2):413–435

Gupta P, Rana KPS, Kumar V, Mishra P, Kumar J, Nair SS (2015) Development of a grey wolf optimizer toolkit in LabVIEW™. In: 2015 international conference on futuristic trends on computational analysis and knowledge management (ABLAZE), Feb 2015. IEEE, pp 107–113

Hao Y, Tian C (2019) A novel two-stage forecasting model based on error factor and ensemble method for multi-step wind power forecasting. Appl Energy 238:368–383

Kamalova A, Navruzov S, Qian D, Lee SG (2019) Multi-robot exploration based on multi-objective grey wolf optimizer. Appl Sci 9(14):2931

Khurma RA, Aljarah I, Sharieh A, Mirjalili S (2020) EvoloPy-FS: an open-source nature-inspired optimization framework in Python for feature selection. Evolutionary machine learning techniques. Springer, Singapore, pp 131–173

Kishor A, Singh PK (2016) Empirical study of grey wolf optimizer. In: Proceedings of fifth international conference on soft computing for problem solving. Springer, pp 1037–1049

Kumar V, Kumar D (2017) An astrophysics-inspired grey wolf algorithm for numerical optimization and its application to engineering design problems. Adv Eng Softw 112:231–254

Kumar V, Chhabra JK, Kumar D (2017) Grey wolf algorithm-based clustering technique. J Intell Syst 26(1):153–168

Lahmiri S (2017) Glioma detection based on multi-fractal features of segmented brain MRI by particle swarm optimization techniques. Biomed Signal Process Control 31:148–155

Li SX, Wang JS (2015) Dynamic modeling of steam condenser and design of PI controller based on grey wolf optimizer. Math Probl Eng 2015:2015

Li Q, Chen H, Huang H, Zhao X, Cai Z, Tong C, Tian X (2017) An enhanced grey wolf optimization based feature selection wrapped kernel extreme learning machine for medical diagnosis. Comput Math Methods Med 2017:1–15

Long W, Xu S (2016) A novel grey wolf optimizer for global optimization problems. In: 2016 IEEE Advanced Information Management, Communicates, Electronic and Automation Control Conference (IMCEC), Oct 2016, IEEE, pp 1266–1270

Long W, Liang X, Cai S, Jiao J, Zhang W (2017) A modified augmented Lagrangian with improved grey wolf optimization to constrained optimization problems. Neural Comput Appl 28(1):421–438

Long W, Jiao J, Liang X, Tang M (2018) An exploration-enhanced grey wolf optimizer to solve high-dimensional numerical optimization. Eng Appl Artif Intell 68:63–80

Luo Q, Zhang S, Li Z, Zhou Y (2016) A novel complex-valued encoding grey wolf optimization algorithm. Algorithms 9(1):4

Mahdad B, Srairi K (2015) Blackout risk prevention in a smart grid based flexible optimal strategy using grey wolf-pattern search algorithms. Energy Convers Manage 98:411–429

Malik MRS, Mohideen ER, Ali L (2015) Weighted distance grey wolf optimizer for global optimization problems. In: 2015 IEEE international conference on computational intelligence and computing research (ICCIC), Dec 2015. IEEE, pp 1–6

Mirjalili S (2015) How effective is the grey wolf optimizer in training multi-layer perceptrons. Appl Intell 43(1):150–161

Mirjalili S, Mirjalili SM, Lewis A (2014a) Let a biogeography-based optimizer train your multi-layer perceptron. Inf Sci 269:188–209

Mirjalili S, Mirjalili SM, Lewis A (2014b) Grey wolf optimizer. Adv Eng Softw 69:46–61

Mirjalili S, Saremi S, Mirjalili SM, Coelho LDS (2016) Multi-objective grey wolf optimizer: a novel algorithm for multi-criterion optimization. Expert Syst Appl 47:106–119

Mittal N, Singh U, Sohi BS (2016) Modified grey wolf optimizer for global engineering optimization. Appl Comput Intell Soft Comput 8(1):1–16

Mohamed AAA, El-Gaafary AA, Mohamed YS, Hemeida AM (2015) Design static VAR compensator controller using artificial neural network optimized by modify grey wolf optimization. Paper presented on 2015 international joint conference on n

Mosavi MR, Khishe M, Ghamgosar A (2016) Classification of sonar data set using neural network trained by gray wolf optimization. Neural Netw World 26(4):393–415

Mustaffa Z, Sulaiman MH, Kahar MNM (2015) LS-SVM hyper-parameters optimization based on GWO algorithm for time series forecasting. In: 2015 4th international conference on software engineering and computer systems (ICSECS), Aug 2015. IEEE, pp 183–188

Niu M, Wang Y, Sun S, Li Y (2016) A novel hybrid decomposition-and-ensemble model based on CEEMD and GWO for short-term PM2.5 concentration forecasting. Atmos Environ 134:168–180

Nur IM, Ülker E (2018) A hybrid cloud-based intrusion detection and response system (IDRS) based on grey wolf optimizer (GWO) and neural network (NN). Paper presented on 2018 international conference on cyber security and computer science (ICONCS'18)

Pan TS, Dao TK, Chu SC (2015) A communication strategy for paralleling grey wolf optimizer. In: International conference on genetic and evolutionary computing, Aug 2015. Springer, Cham, pp 253–262

Parsian A, Ramezani M, Ghadimi N (2017) A hybrid neural network-gray wolf optimization algorithm for melanoma detection. Biomed Res 28(8):3408–3411

Pathak Y, Arya KV, Tiwari S (2019) Feature selection for image steganalysis using levy flight-based grey wolf optimization. Multimed Tools Appl 78(2):1473–1494

Pradhan M, Roy PK, Pal T (2016) Grey wolf optimization applied to economic load dispatch problems. Int J Electr Power Energy Syst 83:325–334

Rajput SS, Bohat VK, Arya KV (2019) Grey wolf optimization algorithm for facial image super-resolution. Appl Intell 49(4):1324–1338

Rodríguez L, Castillo O, Soria J (2016) Grey wolf optimizer with dynamic adaptation of parameters using fuzzy logic. In: 2016 IEEE congress on evolutionary computation (CEC). IEEE, pp 3116–3123

Rodríguez L, Castillo O, Soria J (2017) A study of parameters of the grey wolf optimizer algorithm for dynamic adaptation with fuzzy logic. In: Nature-inspired design of hybrid intelligent systems. Springer, pp 371–390

Saremi S, Mirjalili SZ, Mirjalili SM (2015) Evolutionary population dynamics and grey wolf optimizer. Neural Comput Appl 26(5):1257–1263

Sharma P, Sundaram S, Sharma M, Sharma A, Gupta D (2019) Diagnosis of Parkinson's disease using modified grey wolf optimization. Cogn Syst Res 54:100–115

Shieh CS, Sai VO, Lin YC, Lee TF, Nguyen TT, Le QD (2016) Improved node localization for WSN using heuristic optimization approaches. In: 2016 international conference on networking and network applications (NaNA), July 2016. IEEE, pp 95–98

Singh N, Hachimi H (2018) A new hybrid whale optimizer algorithm with mean strategy of grey wolf optimizer for global optimization. Math Comput Appl 23(1):14

Singh N, Singh SB (2017) Hybrid algorithm of particle swarm optimization and grey wolf optimizer for improving convergence performance. J Appl Math 2017:2017

Sujatha K, Punithavathani DS (2018) Optimized ensemble decision-based multi-focus imagefusion using binary genetic grey-wolf optimizer in camera sensor networks. Multimed Tools Appl 77(2):1735–1759

Sulaiman MH, Mustaffa Z, Mohamed MR, Aliman O (2015) Using the gray wolf optimizer for solving optimal reactive power dispatch problem. Appl Soft Comput 32:286–292

Tu Q, Chen X, Liu X (2019) Multi-strategy ensemble grey wolf optimizer and its application to feature selection. Appl Soft Comput 76:16–30

Turabieh H (2016) A hybrid ANN-GWO algorithm for prediction of heart disease. Am J Oper Res 6(02):136–146

Wang GG, Gandomi AH, Zhao X, Chu HCE (2016) Hybridizing harmony search algorithm with cuckoo search for global numerical optimization. Soft Comput 20(1):273–285

Yadav S, Verma SK, Nagar SK (2016) Optimized PID controller for magnetic levitation system. IFAC-PapersOnLine 49(1):778–782

Yang B, Zhang X, Yu T, Shu H, Fang Z (2017) Grouped grey wolf optimizer for maximum power point tracking of doubly-fed induction generator based wind turbine. Energy Convers Manage 133:427–443

Yogapriya J, Nithya B (2018) Fuzzy based grey wolf optimization for effective medical image retrieval system. In: 2018 international conference on communication and signal processing (ICCSP). IEEE, pp 1046–1050

Zhang S, Zhou Y (2015) Grey wolf optimizer based on Powell local optimization method for clustering analysis. Discrete Dyn Nat Soc 2015

Zhou J, Zhu W, Zheng Y, Li C (2016) Precise equivalent model of small hydro generator cluster and its parameter identification using improved grey wolf optimiser. IET Gener Transm Distrib 10(9):2108–2117

Zhu A, Xu C, Li Z, Wu J, Liu Z (2015) Hybridizing grey wolf optimization with differential evolution for global optimization and test scheduling for 3D stacked SoC. J Syst Eng Electron 26(2):317–328

Chapter 8
A Brief Survey of Neural Networks Trained by the Grey Wolf Optimizer

Abstract This chapter investigates the optimization of Artificial Neural Network (ANN) training through the integration of GWO algorithm. It also identifies various gaps in adoption of GWO while tuning the ANN parameter. Finally, the chapter addressed how GWO can be improved to address those gaps.

Keywords Neural Network (NN) · Grey Wolf Optimizer (GWO) · Optimization · No Free Lunch (NFL) theorem · Convergence · Exploitation · Local optima

8.1 Introduction

Neural Network (NN), which is motivated by biological systems can be applied for information processing effectively. NN has been used extensively over a long period of time due to its dynamic behavior and excellent ability to handle nonlinear data. But, the performance of NN heavily depends on the weights and structure of the network. Moreover, the training of NN is another important issue that needs appropriate algorithm to produce better outcome. Generally, new meta-heuristic algorithms are explored to determine the algorithm's ability to optimize the NN. In this regard, researchers have been investigated recently the optimization ability of NN by GWO.

8.2 Neural Networks Trained by Grey Wolf Optimizer

As a part of investigation to optimize NN applying GWO, Mirjalili (2015) attempted to apply GWO for training multi-layer perceptron. The study gained high exploration and exploitation that could outperform other popular trainers such as PSO, GA, ACO, ES, and PBIL. The investigation is able to produce very competitive results and improve local optima avoidance. The classification accuracy is also very good for the study. However, the study recommended investigating the application of GWO

D. Das et al., *Optimization Algorithms in Machine Learning*, Engineering Optimization: Methods and Applications, https://doi.org/10.1007/978-981-96-3849-9_8

to determine the optimal structure of MLP and fine-tune GWO to produce better solution (Mirjalili 2015).

In another study, Nur and Ülker (2018) investigated the application of GWO for optimizing NN to propose a hybrid cloud-based Intrusion Detection and Response System (IDRS). The study achieved good result, which could successfully detect intrusion over the cloud. Moreover, GWO-NN produced better classification accuracy compared to other classification algorithms such as Naïve Bayes (NB) and Gravitational Search Algorithm with NN (GSA-NN) for two different datasets. However, the classification accuracy of GWO-NN was lower than Multi-layer Perceptron with Back propagation (MLP-BP) for one dataset and lower than both MLP-BP and Particle Swarm Optimization with NN (PSO-NN) for another dataset. In addition, GWO-NN approach was slower in convergence compared to NB during training. Hence, the study recommended modifying GWO to improve the grey wolf performance (Nur and Ülker 2018).

Parsian et al. (2017) attempted to optimize NN applying GWO for melanoma detection. The study trained NN applying GWO to determine the optimal initial weights where the GWO-NN produced better classification rate of 90% compared to ordinary MLP that produced 88%. Moreover, the convergence speed of GWO-NN was really faster and Root Mean Square Error (RMSE) was also reduced through this approach. However, the study did not attempt to compare the performance of GWO-NN with other classification algorithms such as PSO, GA and ACO. Additionally, the investigation did not modify GWO to improve the classification performance, which can be attempted for future research (Parsian et al. 2017).

In another investigation, Turabieh (2016) attempted to optimize NN applying GWO to predict heart disease. The study produced better prediction result for the heart disease related medical dataset through the parameter tuning by GWO-NN and locating initial weights and biases by GWO. Moreover, GWO-NN produced lower RMSE value, i.e., close to 0 and converged much faster. The study also compared the performance of GWO-NN with standard NN and identified that GWO-NN performs much better than standard NN in terms of prediction accuracy, convergence speed and local minima avoidance.

However, the study applied back-propagation algorithm for training and did not attempt to compare the performance of GWO-NN with other classification algorithms such as PSO, GA and ACO. Additionally, the investigation did not modify GWO to improve the classification performance and hence suggested to perform further research to determine the optimal NN structure through the modification of GWO (Turabieh 2016).

Meanwhile, Mosavi et al. (2016) conducted a study to perform classification of sonar dataset applying GWO-NN approach where GWO was implemented for training NN. The investigation was successful to overcome the limitations such as improper classification accuracy, slow convergence speed and trapping in local minimum through GWO-NN. The performance of GWO-NN was also compared in this study with other classification algorithms such as Particle Swarm Optimization (PSO), Gravitational Search Algorithm (GSA), and the hybrid algorithm (i.e. PSOGSA) applying convergence speed, the possibility of trapping in local minimum and classification accuracy metrics for three datasets where, GWO-NN outperformed for all datasets. However, the study did not attempt to modify GWO to investigate the classification performance. Additionally, the study recommended applying GWO or its modification to determine the optimum structure of NN as a future research (Mosavi et al. 2016).

Similar kind of NN optimization through application of GWO was investigated by the authors in Mohamed et al. (2015) for designing of the static VAR compensator (**SVC**) controller for damping power system oscillations. GWO-NN based approach produced better outcome for this research with lower error values and faster convergence. However, the study did not perform the comparison of GWO-NN approach with other classification algorithm such as PSO, GA and ACO. Moreover, modification of GWO could be investigated to produce better classification outcome and determine optimum NN structure (Mohamed et al. 2015).

To sum up, GWO has demonstrated great potential for optimization of NN as a recent swarm intelligence-based meta-heuristic algorithm. However, some limitations of the algorithm needs further investigation such as GWO cannot solve all optimization problems by way of NFL suggestion, GWO can solve only single-objective problems, multi-modal search landscape is difficult to be handled by GWO because the operators are converged to identical solution, more number of variables worsens the performance of GWO due to entrapment in local solutions, GWO may produce local solutions for a problem containing large number of variables and local solutions due to faster convergence and exploitation, GWO has also limitation in terms of exploration rate as it has the possibility of being stagnant with its limited operators' alpha, beta and delta.

Moreover, the encircling model recommended by GWO may performs the exploration to limited extent only, so GWO needs more operators to increase the exploration rate. Hence, GWO needs to be modified or extended to solve complex problems (Faris et al. 2018; Gupta and Deep 2017; Mirjalili 2015; Mirjalili et al. 2014; Nur and Ülker 2018).

8.3 Gap Analysis on the Need for Modified Grey Wolf Optimizer

Table 8.1 presents the gap analysis in the current adoption of GWO for tuning ANN where the common approach GWO is applied to optimize ANN for different type of investigations and datasets.

As seen from the analysis, the demand for proposing a new algorithm or improving earlier algorithm is enormous to deal with the limitation of existing algorithms. The requirement of a new optimization algorithm is also highlighted by No Free Lunch (NFL) theorem that a single algorithm cannot solve all the optimization problems optimally (Wolpert and McReady 1997). In line with the gap analysis discussed earlier, this research will endeavor to address the gap indicated above by planning

Table 8.1 Gap analysis findings summary

Strategy	Main features	Limitations and recommendations
GWO-MLP (Mirjalili 2015)	• Addressed the exploration and exploitation • Better performance than popular trainer such as PSO, GA, ACO, ES and PBIL • Improved local optima avoidance • Good classification accuracy for selected dataset	• Investigation is required to apply GWO for determining optimal structure of MLP • Fine tuning of GWO is required to produce better solution
GWO-NN (Mosavi et al. 2016)	• The investigation addresses the limitations such as improper classification accuracy, slow convergence speed and trapping in local minimum • Good performance for selected datasets	• Classification performance is not at 95% confidence level for different dataset • Modification of GWO is required for different dataset to enhance performance
GWO-NN (Turabieh 2016)	• Produced good prediction result for the heart disease related medical dataset • Produced lower RMSE value • Better performance than standard NN	• Classification performance is not at same level for other dataset • GWO needs to be modified to improve the classification performance for different dataset
GWO-NN (Parsian et al. 2017)	• Produced better classification rate compared to ordinary MLP • Faster convergence • Reduced Root Mean Square Error (RMSE)	• Classification performance is not at 95% confidence level for various datasets • Modification of GWO is recommended for better performance
GWO-NN (Nur and Ülker 2018)	• Able to detect intrusion over the cloud • Balanced exploration and exploitation • Better classification accuracy compared to Naïve Bayes (NB) and Gravitational Search Algorithm (GSA)	• Classification accuracy is lower for some dataset • Slower in convergence compared to NB • Modification of GWO is suggested

and executing another attempt in view of modified Grey Wolf Optimizer (MGWO) applying ensemble approach.

Grey Wolf Optimizer (GWO) is one of the most recent swarm intelligence-based meta-heuristic algorithms shaped for addressing the problem of global optimization. Grey wolves' hunting and leadership hierarchy in nature motivates the inspiration of such algorithm. In light of supplementing existing work on meta-heuristic based ensemble strategies, adopting Modified Grey Wolf Optimizer (MGWO) has all the earmarks of being an appealing choice. Specially, GWO has benefits over other meta-heuristic algorithms (Faris et al. 2018; Gupta and Deep 2017; Mirjalili et al. 2014):

(i) GWO is simple and flexible SI-based algorithm that produces random population of grey wolves. The computation facilitated by GWO is lightweight compared to other meta-heuristic algorithm like GA and PSO.
(ii) GWO implements intense activities controlled by two parameters to balance exploration and exploitation so that local optima stagnation can be avoided.
(iii) The mathematical model offered by GWO is novel, although the estimation of global optimum is analogous to other population-based algorithm. Moreover, GWO has the ability to displace a solution to another n-dimensional search space.
(iv) GWO requires less memory contrasted with PSO as it contains only one vector because, PSO requires two vectors namely, position and velocity. Additionally, GWO retains just three best solutions, while PSO retains one best solution gained through all particles. The mathematical calculations of PSO and GWO are dissimilar. GWO is considered as a standout among the most developing SI algorithms. The success of GWO algorithm propels different scientists to apply the algorithm for various optimization problems. Till date, GWO has been utilized successfully for solving numerous problems but not limited to, global optimization problems, electric and power engineering problems, scheduling problems, power dispatch problems, control engineering problems, robotics and path planning problems, environmental planning problems (Faris et al. 2018).

However, GWO can be improved to address few shortcomings:

(i) Balancing of convergence and exploitation is required for GWO to avoid local optima. Because, the current best optimal individual is biased toward alpha and other individuals (e.g., beta and delta) attempt to modify their positions toward this best individual in each iteration process. Consequently, this update process may cause the algorithm to fall to local optima especially in the cases where there are many competing local optima.
(ii) Poor search agents among alpha, beta, delta wolves of GWO deteriorate exploitation. Hence, the exploitation needs to be improved by avoiding such search agents.
(iii) Poor search agents around search space worsen exploration. So, the exploration is required to be enriched by avoiding similar search agents.
(iv) The positions of wolves are updated mostly based on the experience of alpha, beta, and delta leaders in GWO that lead to premature convergence. Thus, measure should be taken to keep away premature convergence.
(v) More number of variables degrade the performance of the GWO algorithm due to the entrapment of the initial population in a local solution. So, number of variables need to be controlled.

8.4 Summary

This chapter presents the explanation of various Neural Networks trained by GWO-based algorithms. The chapter investigated that there are still limitations with existing meta-heuristic based models in terms of entrapment in local minima, balance between exploration and exploitation, and convergence. The next chapter will devise an MGWO based ensemble model for stock prediction to address the gaps persist with existing model.

References

Faris H, Aljarah I, Al-Betar MA, Mirjalili S (2018) Grey wolf optimizer: a review of recent variants and applications. Neural Comput Appl 30(2):413–435

Gupta S, Deep K (2017) Performance of grey wolf optimizer on large scale problems. AIP Conf Proc 1802(1):020005

Mirjalili S (2015) How effective is the grey wolf optimizer in training multi-layer perceptrons. Appl Intell 43(1):150–161

Mirjalili S, Mirjalili SM, Lewis A (2014) Grey wolf optimizer. Adv Eng Softw 69:46–61

Mohamed AAA, El-Gaafary AA, Mohamed YS, Hemeida AM (2015) Design static VAR compensator controller using artificial neural network optimized by modify grey wolf optimization. Paper presented on 2015 international joint conference on neural networks

Mosavi MR, Khishe M, Ghamgosar A (2016) Classification of sonar data set using neural network trained by gray wolf optimization. Neural Netw World 26(4):393–415

Nur IM, Ülker E (2018) A hybrid cloud-based intrusion detection and response system (IDRS) based on grey wolf optimizer (GWO) and neural network (NN). Paper presented on 2018 international conference on cyber security and computer science (ICONCS'18)

Parsian A, Ramezani M, Ghadimi N (2017) A hybrid neural network-gray wolf optimization algorithm for melanoma detection. Biomed Res 28(8):3408–3411

Turabieh H (2016) A hybrid ANN-GWO algorithm for prediction of heart disease. Am J Oper Res 6(02):136–146

Wolpert DH, McReady WG (1997) No free lunch theorems for optimization. IEEE Trans Evol Comput 1:67–82

Chapter 9
Improving the Exploration and Exploitation of Grey Wolf Optimizer

Abstract This chapter presents the formation of Modified Grey Wolf Optimizer (MGWO) by reviewing meta-heuristic algorithms, neural network, ensemble model, and existing model including the strengths and limitations of GWO in the previous chapter. It also explains the variants of GWO. Besides, various parameters of MGWO, exploration and exploitation through MGWO as well we formation of mathematical models of MGWO are demonstrated in this chapter.

Keywords Grey Wolf Optimizer (GWO) · Modified Grey Wolf Optimizer (MGWO) · Exploration and exploitation in MGWO · Exploration · Exploitation · Mathematical model of MGWO

9.1 Introduction

In the previous chapter, meta-heuristic algorithms, neural network, ensemble model, and existing model were reviewed. The chapter also highlighted the research gap, the strength and limitation of existing meta-heuristic algorithms including Grey Wolf Optimizer (GWO).

Proceeding from the preceding chapter, this chapter portrays the research methodology applied for designing, implementing and assessing the ensemble approach consisting Modified Grey Wolf Optimizer (MGWO) and neural network. This chapter likewise depicts tuning of MGWO's exploration to accomplish the optimal outcome. The chapter also illuminates the three research phases to fulfill the aims and objectives of the research namely k-means clustering for categorizing the stock data, classification to determine the best features applying meta-heuristic algorithm and neural network model to predict the stock price.

In this chapter, GWO is modified and enhanced to better suit with stock data. The main focus here is to develop an ensemble model through the review and enhancement of algorithm such as clustering, classification and prediction. The chapter further demonstrates the neural network architecture, experimentations, and validation of the

D. Das et al., *Optimization Algorithms in Machine Learning*, Engineering Optimization: Methods and Applications, https://doi.org/10.1007/978-981-96-3849-9_9

constructed neural network. Data collection method and preprocessing for k-means clustering, neural network, and MGWO is also elaborated in this chapter.

9.2 The Proposed Variant of Grey Wolf Optimizer

The overall research plan is presented in Fig. 9.1. The first section of the figure represents the detail discussion about GWO and Modified GWO. Where, the basic characteristics of original GWO, mathematical model, exploration and exploitation through GWO will be discussed initially. Then, the modification of GWO, parameters of the algorithm, mathematical model, balance of exploration and exploitation through Modified GWO will be presented. The ensemble model applying neural network and Modified GWO is discussed in the next section, where data preprocessing and the feature selection will be explained as well. Next, application of ensemble model and evaluation will be presented. Finally, the performance evaluation of the ensemble model will be discussed.

Figure 9.2 portrays the research activities for design and application of Modified Grey Wolf Optimizer (MGWO) where, the MGWO is designed by studying the organization and movement of Grey Wolf from literature besides the development of Grey Wolf based mathematical model. The MGWO is a population-based meta-heuristic algorithm where the solution of a problem is produced through the movement of agents by balancing the exploration and exploitation to the search space. The proposed algorithm mimics the behavior of Grey Wolf in this research, where

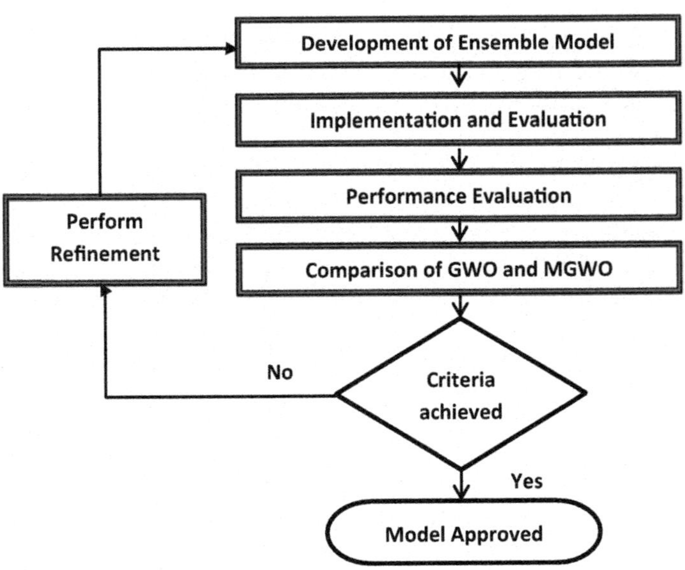

Fig. 9.1 Overall research plan

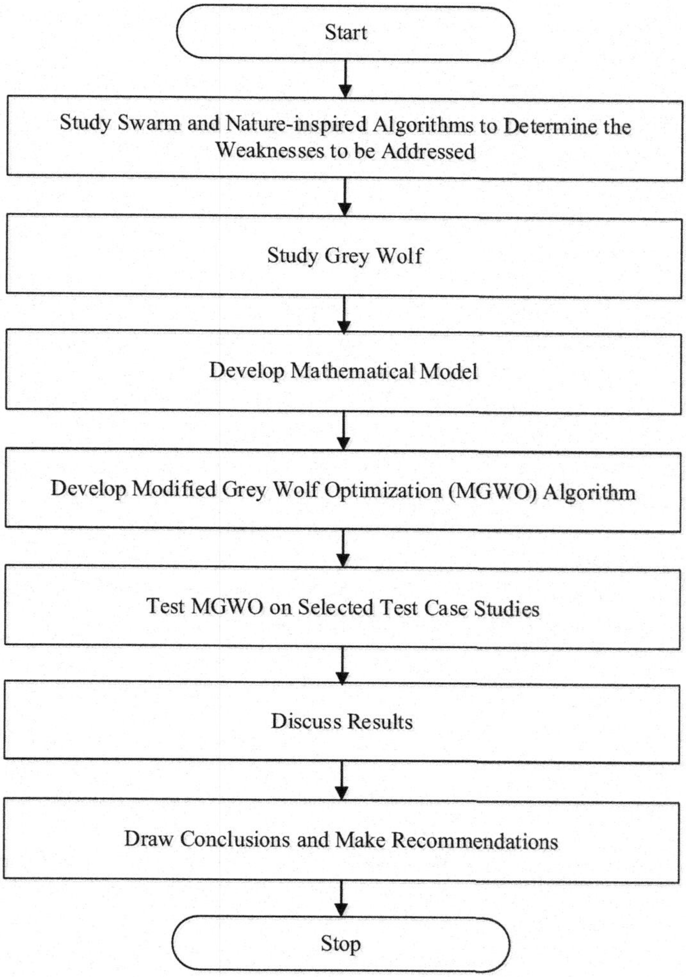

Fig. 9.2 MGWO research activities flowchart

the algorithm is capable to balance the exploration and exploitation to the search space as it simulates the democratic and communicative behavior of Grey Wolf for solutions to their search.

9.3 Modified Grey Wolf Optimizer Algorithm

This section illustrates the design of proposed algorithm, MGWO. In the current research, MGWO is an approach based on GWO for feature selection and prediction. Original GWO starts by initializing the grey wolf population. Later, the algorithm updates the position of each search agent by the locations of the best solutions, and assessed the objective function of the algorithm.

The proposed algorithm improves the wolves attack strategy. It calculates the weights based on wolves fitness function and gives the highest weight to the dominant wolf concurrently to improve the convergence, decide the suitable thresholds faster, and provides good classification rate, efficiency and accuracy. Table 9.1 describes the definition of each MGWO's parameter where, the parameter value for maximum iteration and population size are maintained with GWO and other algorithm for fair comparison.

The value for coefficient vectors $\vec{a}$, $\vec{A}$, and $\vec{C}$ are reduced here in MGWO from original GWO algorithm in order to ensure closer exploration and exploitation. As the probability value lies between 0 and 1, the value for Probability vector is chosen as 0–1. Finally, the probability of generating 1 or 0 for each iteration is equal and hence, the Threshold value is chosen as 0.5 (Monteiro et al. 2018). The complete pseudocode of MGWO algorithm can be represented in Fig. 9.3.

In pseudocode of MGWO illustrated as Fig. 9.3, *p_size* represents the grey wolf population to initialize the MGWO parameters; max_iter denotes maximum number of iteration. The initial positions of grey wolves are generated at this stage. The initialization of an archive (*A*) will be made to collect the solution for each iteration. Then, the initialization of the vectors *a, A, C* are performed through Eqs. (9.1) and (9.2).

$$\vec{A} = 2\vec{a} \cdot \vec{r}_1 - \vec{a} \tag{9.1}$$

$$\vec{C} = 2 \cdot \vec{r}_2 \tag{9.2}$$

The values of $\vec{a}$ are linearly decreased from 1 to 0 over the course of iterations as per the original algorithm. The initialization of the controlling parameter for selecting or removing solutions from archive (*A*) is made where (Selecting $G^n_\alpha P_1 = 1/N_i$ and Removing $G^n_\alpha P_2 = N_i$). Then, the estimation of the fitness for each search agents is made and best hunt agents are identified where, the best hunt agent is G_α, the second

Table 9.1 Details of MGWO's parameters

Parameter	Values
Maximum number of iterations (max_iter)	300
Population size (*p_size*)	1000
$\vec{a}$	Linearly decreased from 1 to 0
$\vec{A}$	Random values in $-a$ to a
$\vec{C}$	Random values in 0–1
Probability vector (*P*)	0–1
Threshold	0.5
Archive (*A*)	Collected solution in each iteration

```
Begin
Initialize p_size, max_iter, n, pos, flag
Generate the init_pos of grey wolves randomly
Construct A archive of collected solutions in each iteration
Initialize a, A⃗ and C⃗
Initialize Controlling parameter of selecting/removing solutions from archive A
(Selecting Gα^n P1=1/Ni , Removing Gα^n P2 = Ni)
Gα= The grey wolf with the first highest fitness
Gβ= The grey wolf with the second highest fitness
Gδ= The grey wolf with the third highest fitness
Threshold = 0.5
P(Gα^n) = 1, the probability vector, (Π = {Gα^1, Gα^2, Gα^3, …………, Gα^n})
While i<max_iter
  Calculate the fitness of grey wolves
  If fitness i_th < Gα
     Update Gα with new fitness i_th value
     Update Gα Position
  Else If fitness i_th > Gα and fitness i_th < Gβ
     Update Gβ with new fitness i_th value
     Update Gβ Position
  Else If fitness i_th > Gα and fitness i_th > Gβ and i_th <Gδ
     Update Gδ with new fitness i_th value
     Update Gδ Position
  End If
  For x = 1:p_size
    For y = 1:n
      If pos(x,y) > Threshold
        flag(y) = 1
      Else
        flag(y) = 0
      End If
    End For
  End For
  Update the position of current grey wolf by, G⃗(i + 1) = ( G⃗1 + G⃗2 + G⃗3 ) / 3
  Calculate the probability, P(Gα^n) = N(Gα^n, A) / Σ_{j=1}^{n} N(Gα^j, A)
  If P(Gα^n)<Threshold
    v=abs(Max (Gα) − fitness_value)/fitness_value   // Normalization
    For z = 1: p_size
       Find non-dominated P1(Gα)< v
      Update the archive A, remove Gα_th
     Re-Generate the init_pos of Gα_th randomly
   End For
  Else
   Update a, A⃗ and C⃗
   Calculate the fitness of grey wolves including selected features
   Update Gα, Gβ, and Gδ
  End If
  i = i + 1
End While
Return Gα, selected features
End
```

Fig. 9.3 Pseudocode of MGWO algorithm

best hunt agent is G_β and the third best hunt agent is G_δ. Here, the initialization of the Threshold value is assigned to 0.5 as each of the steps may produce different values ranging between 0 and 1. The reason for choosing Threshold value 0.5 is that the probability of generating 1 or 0 for each iteration is equal for such value (Monteiro et al. 2018).

Hence, the probability for all best search agent (G_α) is initialized to 1. Then, the location of the hunt agents are updated based on fitness. Here, the position of the agent is checked in comparison with threshold value 0.5, if the threshold value is more than 0.5, then feature will be selected by updating flag value to 1 otherwise feature will not be selected and updating flag value to 0. The location of the current hunt agent is updated using Equation, $\vec{G}(i+1) = \left(\vec{G}_1 + \vec{G}_2 + \vec{G}_3\right)/3$. The probability for all the best search agent is calculated applying the equation, $P\left(G_\alpha^n\right) = \frac{N\left(G_\alpha^n, A\right)}{\sum_{j=1}^{n} N\left(G_\alpha^j, A\right)}$.

Then, the probability of best search agent will be compared with threshold, if the probability value is below threshold then the non-dominated values will be removed from the archive and others will remain with archive for re-generation, otherwise the vectors a, $\vec{A}$, and $\vec{C}$ will be updated, the fitness value of all hunts including selected features will be estimated and the value of search agents G_α, G_β and G_δ will be updated. Hence, the stopping condition will be checked whether the iteration (i) reaches max number of iterations. Finally, the best value of solution G_α will be returned with selected features.

9.4 Exploration and Exploitation in MGWO

As the algorithm needs to address the exploration and exploitation, MGWO is tuned to balance the exploration and exploitation. To emphasize the exploration, the algorithm determines the new area in the problem search space through the application of rapid alteration in the solution so that the algorithm does not stack in local minimum. The exploitation is balanced by improving the accomplished expected solution in exploration by determining the neighborhood of every solution.

In MGWO, $\vec{C}$ and $\vec{A}$ are the main controlling parameter to ensure exploration by returning a random values between 0 to 1 for $\vec{C}$ and 1 to 0 linearly decreased value of a for $\vec{A}$. Whereas, the original GWO, promotes the exploration through larger range of random values between 0 to 2 for $\vec{C}$ and 2 to 0 linearly decreased value of a for $\vec{A}$. The value of parameter $\vec{A}$ ranges between -1 and 1 where, exploration is achieved through the value $A < 0$ and exploitation is achieved through the value $A > 0$. The balance between exploration and exploitation is maintained through this algorithm by setting random values for $\vec{C}$ and linearly decreased values for $\vec{A}$.

Moreover, original GWO searches for best G_α to emphasize the exploration and exploitation with higher ranges of values whereas, MGWO attempts to determine multiple G_α to explore and exploit with relatively shorter range of values. If unsuccessful with one G_α, the algorithm proceeds to find another G_α to further explore and exploit. This way, MGWO makes a good balance between the local and global solutions which also balances the exploration and exploitation.

9.5 Mathematical Model of MGWO

The mathematical model of MGWO can be formed as per the equations below:

$$\vec{D} = \left|\vec{C} \cdot \vec{G}_a(i) - \vec{G}(i)\right| \tag{9.3}$$

$$\vec{G}(i+1) = \vec{G}_a(i) - \vec{A} \cdot \vec{D}, \tag{9.4}$$

where, i denotes the current iteration, $\vec{A}$ and $\vec{C}$ are coefficient vectors, $\vec{G}_a$ is the position vector of the prey, and $\vec{G}$ denotes the position vector of a grey wolf.

The vectors $\vec{A}$ and $\vec{C}$ are calculated as follows:

$$\vec{A} = 2\vec{a} \cdot \vec{r}_1 - \vec{a} \tag{9.5}$$

$$\vec{C} = 2 \cdot \vec{r}_2, \tag{9.6}$$

where components of $\vec{a}$ are linearly reduced from 1 to 0 over the number of iterations and is used for controlling the trade-off between exploitation and exploration. The following equations will be employed for updating the value of variable:

$$\vec{a} = 2 - i(2/X_i) \tag{9.7}$$

$$\vec{G}(i+1) = \left(\vec{G}_1 + \vec{G}_2 + \vec{G}_3\right)/3, \tag{9.8}$$

where, X_i denotes the number of iterations, $\vec{r}_1$ and $\vec{r}_2$ are random vectors between [0, 1] which are employed to find the optimal solution. Appropriate idea about the potential location of prey can be availed by Alpha, Beta and Delta, where they help the Omega to follow the suitable positions. The values of $\vec{G}_1$, $\vec{G}_2$ and $\vec{G}_3$ can be obtained through the equations below:

$$\vec{G}_1 = \vec{G}_\alpha - \vec{A}_1 \cdot \vec{D}_\alpha \tag{9.9}$$

$$\vec{G}_2 = \vec{G}_\beta - \vec{A}_2 \cdot \vec{D}_\beta \tag{9.10}$$

$$\vec{G}_3 = \vec{G}_\delta - \vec{A}_3 \cdot \vec{D}_\delta \tag{9.11}$$

In iteration i, the best 3 solutions are, respectively, $\vec{G}_1$, $\vec{G}_2$, and $\vec{G}_3$. Where, the values of $\vec{D}_\alpha$, $\vec{D}_\beta$ and $\vec{D}_\delta$ are as indicated in Eqs. (9.12), (9.13) and (9.14):

$$\vec{D}_\alpha = \left|\vec{C}_1 \cdot \vec{G}_\alpha - \vec{G}\right| \tag{9.12}$$

$$\vec{D}_\beta = \left|\vec{C}_2 \cdot \vec{G}_\beta - \vec{G}\right| \tag{9.13}$$

$$\vec{D}_\delta = \left|\vec{C}_3 \cdot \vec{G}_\delta - \vec{G}\right| \tag{9.14}$$

Here, n random parameter vectors are formed to further explore the best solution for every iteration to be selected on, to update the vectors as per Eq. (9.15).

$$\text{Vectors},\ \Pi = \left\{G_\alpha^1, G_\alpha^2, G_\alpha^3, \ldots G_\alpha^n\right\}. \tag{9.15}$$

If N is average distance between wolves, A is shared archived score and G_α^j is sum of all the best solution then the following probability Eq. (9.16) can be formed as per meta-population distribution.

$$\text{Probability},\ P\left(G_\alpha^n\right) = \frac{N\left(G_\alpha^n, A\right)}{\sum_{j=1}^n N\left(G_\alpha^j, A\right)} \tag{9.16}$$

In MGWO, two important steps are included that comprises: firstly, select the features and train the neural network using MGWO to determine the optimal initial weights and secondly, test the results of the proposed MGWO approach. This approach can improve the efficiency of the back-propagation to seek global optima in the search space. For the proposed MGWO approach, the weights are achieved as a vector of variables.

For this approach, Root-Mean-Square Error (RMSE) is the cost function which can determine the error between the actual value and predicted value which can be expressed by Eq. (9.17):

$$\text{RMSE} = \sqrt{\sum_{i=1}^{n} (a_i - p_i)^2}, \tag{9.17}$$

where, n is the number of observations, a_i is the number of actual values and p_i is the number of predicted values from neural network. The lower RMSE value is expected for determining acceptable prediction and making the model acceptable.

9.6 Summary

In this chapter, the detailed design of the modified MGWO along with the ensemble model are presented and discussed. The main key improvement of GWO was also discussed with the current limitation it. Sections 9.3, 9.4, and 9.5 have discussed the detailed technical side of the modifications were applied to achieve the MGWO algorithm along with its mathematical modeling. In the next chapter, the use of developed ensemble model in predicting stock market data will be further discussed.

Reference

Monteiro C, Ramirez-Rosado IJ, Fernandez-Jimenez LA, Ribeiro M (2018) New Probabilistic Price Forecasting Models: Application to the Iberian Electricity Market. Int J Electr Power Energy Syst 103:483–496

Chapter 10
How to Predict Stock Market Using Neural Networks and Grey Wolf Optimizer

Abstract This chapter starts with the review of stock market. Then, the formation of Ensemble algorithm consisting of ANN and MGWO as well as its implementation for feature selection and stock prediction are demonstrated here. A flowchart illustrates in this chapter the feature selection and training of ANN. The chapter also evaluates the effectiveness of Ensemble algorithm for stock prediction. Finally, the chapter concludes with the discussion about the computational complexity of MGWO. Overall, this chapter detail investigates how the application of Ensemble model with ANN and MGWO affects the stock prediction.

Keywords Stock market prediction · Modified Grey Wolf Optimizer (MGWO) · K-means clustering algorithm · Data partitioning · Evaluation of predicted stock price · Feature selection · Ensemble of MGWO · Computation complexity of MGWO

10.1 Introduction

Stock market remains best investment alternatives for few decades despite being unpredictable and uncertain. The economy of a country can be greatly affected by stock market as it plays a significant role to the economy. Investors invest in the stock market to acquire the profit and for that they purchase the security bond of different company. The selection of the security bond of different company is made based on the different factors such as company's information analysis and prediction, and dividend declaration.

If emerging stock market is considered as an example, it is observed that, most investors do not have adequate information about the market analysis and prediction of the future prices. Investors purchase the security bond based on rumor, manipulated financial report of companies and without any idea about data analysis and prediction. As a result, the stock market becomes unpredictable due to extreme ups and downs in the daily share price indices. Investors lose their capital in the unstable stock market, which creates a big crisis in the capital market, and national economic growth is

D. Das et al., *Optimization Algorithms in Machine Learning*, Engineering Optimization: Methods and Applications, https://doi.org/10.1007/978-981-96-3849-9_10

greatly hampered due to such crisis. Therefore, a good model is required that will provide real scenario of stock market and facilitate the investors to predict the prices in advance. All these will contribute toward the solidity of the national economy.

10.2 Stock Market Prediction

Despite prediction of stock price is extremely complicated due to the nonlinear form of stock data, stock price prediction is an exciting research area. Consequently, researchers are continuously striving to improve the existing prediction models. Individual and institutional investors are not leaving even single efforts to make an accurate stock investment plan. They are devising their own strategy to perform the daily and future investment.

However, due to the complex nature of stock data and stock market, stock price prediction still remains one of the most complicated jobs in financial forecasting (Wei 2013). Investors are grabbing any forecasting method that assists them in making decent profit and minimize investment risk through stock investment.

Consequently, it enables researchers' abundant motivation to either develop a new or enhance various stock prediction models (Atsalakis et al. 2011). Different types of prediction models have proved to be effective for stock market as the investors can avail the profit through those stock prediction models. Neural network is widely used by many researchers due to its ability to learn from unknown hidden patterns and capability to produce solution from unknown data. Some stock prediction works related to neural network are included here. ANN and ARIMA models are used for forecasting next day stock market by Merh et al. (2011). Future index value of Sensex (BSE 30) was also forecasted by them through those models to determine the forecasting accuracy.

Mahajan et al. (2015) proposed a Neuro-Fuzzy model for BSE India which could guide investors to have profitable script in their portfolio. However, integration of multiple approaches are gaining priority currently instead of single approach in order to improve the stock price prediction model where distinct feature of each model is combined together to build a rigorous stock prediction model (Wang et al. 2012). Due to unpredictable behavior of stock market there is always some risk involved to the investment (Hassan and Nath 2005). Moreover, stock prediction is even more complex due to influence of different factors: positive or negative news of the company, political turbulences, rate of interest, dollar price, and natural disasters (Bonde and Khaled 2012).

Random Walk Hypothesis (RWH) believes that stock price is not affected by historical price and tomorrow's price is predictable through the analysis of today's price. The researchers who support RWH also established that stock prices cannot be predicted as they follow random behavior and it is unnecessary to apply fundamental analysis or machine learning for predicting stock market. Efficient Market Hypothesis (EMH) is another divisive model that also explains RWH states that stock price of a security is the reflection and determination of all relevant information.

According to the EMH, buying stock is a game of chance and hence investors may not be able to analyze the information with better efficiency. In spite of controversy with EMH, researchers progress the stock prediction research forward through numerous research publications in this area (Tilakaratne et al. 2009). The RWH also suggests that stock data do not follow patterns, and is therefore not eligible for prediction. Extensive research on the topic implies the opposite that technical analysis can produce positive results in terms of prediction (Wong et al. 2012). Research on reward of technical analysis on the Singapore market suggests that a significant part of member companies rely on technical analysis (Wong et al. 2012).

The behavior of stock market may not be well known by financial analysts and eventually they will not be able to judge the exact time to buy or sell stocks for making profit through stock investment. However, decision making is a critical and vital process in stock trading as it has to be made correctly and at the right time (Gamil et al. 2007).

Due to higher profit through stock investment, stock exchange is a prevalent investment destination though recent experience has demonstrated that the higher the expected return, the greater the risk consequences (Kuo et al. 2001; Vincent and Bamiro 2013). Thus, various studies have led to different decision support models in order to provide investors with optimal prediction. Internet plays vital role to make the huge stock information available to the investors, but the investor's tasks become quite tough due to various responsibilities such as collection, analysis, filtration and making correct decision from several information (Van den Berg et al. 2004).

Different stock prediction models have been developed over the years to understand, monitor and predict the stock market worldwide. The applications of various artificial intelligence-based models to the stock market has drawn the researchers' attention apart from the statistical models that have been used to understand and predict fluctuations in the stock market. Many researchers have also focused on technical analysis as the procedure to improve the investment rate in stock market.

Kozdraj (2009) attempted to apply neural network in predicting stock price for Warsaw Stock Exchange, Wu and Coggeshall (2012) applied neural network for the purpose of training multilayer perceptron network, Lopez et al. (2012) performed classification of data to build a model through placing similar data in a same group whereas disparate data is separated through clustering, Lertyingyod and Benjamas (2016) proposed a stock prediction model through the analysis of historical price of stock and applied Data Mining techniques to predict one, five and ten day periods stock price trend, Narayanan and Govindarajan (2015) applied combination of classification model applying SVM and Naïve Bayes which gave them more accuracy with significant reduce of classification error, Navale et al. (2016) applied Artificial Neural Network (ANN) for prediction and Patel et al. (2015) applied SVM for stock price prediction. Therefore, understanding the market and being able to predict what will happen at the near future are desirable skills for every investor.

Each stock market adopts unique characteristics and the information gained from one perhaps implemented in another. Hence, the unique features of stock market need to be studied through research. Despite neural network proved its potential in modeling nonlinear relationship of stock market, adopting it for stock market is still challenging (Lertyingyod and Benjamas 2016; Navale et al. 2016; Narayanan and Govindarajan 2015; Patel et al. 2015). The challenges through the neural network includes determining appropriate neural network architecture, the selection of representative input vectors of features from the time series data of the market and the availability of sufficient data for training. Stock market has adopted various algorithms for prediction over the years, i.e., GA, PSO, and SVM. The important question may raise as to, which model is the most effective for stock prediction?

Neural network and similar models are applied extensively for stock prediction by most studies. Extensive domain knowledge is the utmost priority in this regard to determine an appropriate choice of data and models for building neural network based stock prediction model. In spite of abundant investigations toward design and development of the standard model for stock prediction, no generic stock prediction model has been revealed yet (Lahmiri and Boukadoum 2015).

Moreover, many investigators have developed several models in neural network to predict stock market, but most has failed to provide appropriate prediction due to various issues such as entrapment in local minima, result inaccuracy, slow convergence rate, uncertain and instable market situations. Researchers investigated various heuristic models over the years for training neural network to propose a good stock prediction model such as PSO, GA, ACO, ES, and PBIL.

From the above discussions, it is clear that further investigations can be attempted to implement a good prediction model for stock market. Hence, the present study selects stock prediction as a case study for the implementation of ensemble model so that an effective stock prediction approach can be proposed that can predict average daily price of listed companies for stock market.

Figure 10.1 presents the block diagram of the activities for stock prediction where, k-means clustering algorithm is applied in the pre-processing block for selection of organization based on growth. However, the pre-processing block will be briefly explained in this research. The main contribution of the research is stock prediction and evaluation block and hence, the block will be explained in detail. The prediction of stock market applying ensemble model consists of the steps:

1. Select and preprocess datasets
2. Create a data mining based decision support model applying k-means clustering to categorize the organization based on growth
3. Create a classification based decision support model which can evaluate the suitability of data, select the features and train the neural network for prediction
4. Create an MLP neural network based decision support model to forecast the stock price
5. Create and benchmark a comparison strategy to evaluate the performance of the prediction model.

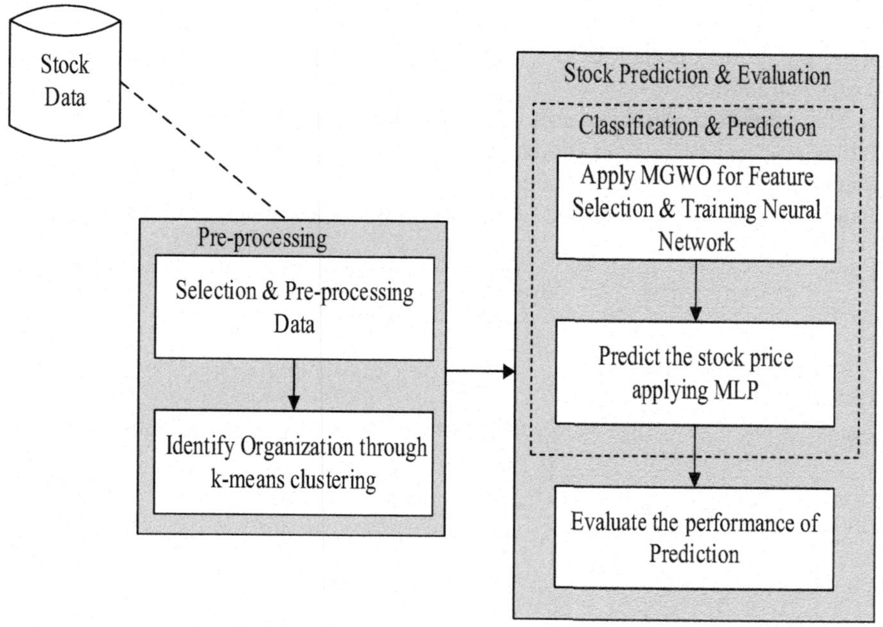

Fig. 10.1 Block diagram of stock prediction activities

10.3 MGWO Implementation

MGWO consists of three basic operations such as initialization; calculation of fitness value, update accordingly and determine the best position of the grey wolf; and finally, validation of the terminating conditions.

10.3.1 Data Collection and Analysis

This study uses 6 years of end-of-day data beginning in January 2011. The research will apply the collected data to evaluate the proposed stock prediction ensemble model consisting of neural network and MGWO. The proposed model is evaluated in the context of the stock markets NYSE, NASDAQ, Bursa Malaysia, and DSE, Bangladesh. Therefore most of the analyzed data originates from the Yahoo Finance, Bursa Malaysia Library and DSE Library. The factors data is collected from Bangladesh Bank Website, Bangladesh Jewelry Samity and Banks in Bangladesh. In the following, the data source, pre-processing steps and arrangement of datasets are described in detail. All data is preprocessed and back adjusted as per requirement.

Table 10.1 shows typical examples of stock dataset from DSE, Bangladesh, where dataset belongs to one particular company of pharmaceutical sector named ACI for a certain date as mentioned in the table. Table 10.2 shows factors dataset used for

DSE, Bangladesh, where dataset belongs to the price for the factors such as gold price (per grams), dollar price (1 unit), bank interest rate, foreign direct investment (FDI), and inflation.

The dataset is preprocessed to prepare it for experimentation. As observed in Tables 10.1 and 10.2, some numbers are very big in range. So, min–max conversion is applied to scale the data to same range as per Eq. (10.1) where, N_i is normalized data, y_i is original data for $i = 1, 2, 3, \ldots, n$ and n is total number of observations.

$$N_i = \frac{y_i - \min(y)}{\max(y) - \min(y)} \tag{10.1}$$

Table 10.1 Stock dataset from DSE

Date	Trading code	LTP*	High	Low	OPENP*	CLOSEP*	YCP	Trade	Value (mn)	Volume
1/1/2017	ACI	387	390	386.2	388	386.7	385.5	603	25.792	66,644
1/2/2017	ACI	403	404	390	390	402.1	386.7	1431	79.082	199,196
1/3/2017	ACI	415.2	415.8	402	402	414.7	402.1	1685	87.066	212,297
1/4/2017	ACI	413.5	422.9	412.8	418.6	413.5	414.7	1088	62.533	150,348
1/5/2017	ACI	408.5	416	404	416	406.4	413.5	941	49.441	120,674
1/8/2017	ACI	407	414.4	402.8	414.4	408.8	406.4	816	46.914	115,185
1/9/2017	ACI	414	418.9	405.3	410	415.6	408.8	1238	73.354	177,217
1/10/2017	ACI	432	432	416.1	418.3	430.1	415.6	1825	111.914	264,037
1/11/2017	ACI	433	437.6	430	434.6	431.6	430.1	1348	76.027	175,308
1/12/2017	ACI	424.1	432	421	431.6	423.3	431.6	950	46.643	109,565
1/15/2017	ACI	430	431	424.1	424.1	427.6	423.3	903	48.854	114,000
1/16/2017	ACI	428.7	437	426.2	429.8	430.2	427.6	1040	65.912	152,235
1/17/2017	ACI	427	431	426.5	430.9	427.7	430.2	903	48.841	113,897
1/18/2017	ACI	423.5	430	423	427.1	424.4	427.7	864	52.773	123,756
1/19/2017	ACI	421	428	417.1	423	421.5	424.4	817	57.738	136,874

Table 10.2 Factors dataset used for DSE

Gold price	Dollar price	Bank interest rate	FDI	Inflation
2905	69.25	8.50	861,736,237.16	9
2995	69.25	8.50	861,736,237.16	9
3095	69.24	8.50	1,184,776,059.05	8
3190	69.24	10	1,184,776,059.05	8
3290	69.24	10	1,474,542,605.46	7
3190	69.27	10	1,474,542,605.46	7
3260	69.26	12	1,474,542,605.46	9
3340	69.25	12	1,501,647,072.05	9
3430	69.26	12.50	1,501,647,072.05	8
3455	69.26	12.50	1,501,647,072.05	11
3535	69.25	12.50	1,501,647,072.05	11

10.3.2 K-Means Clustering Algorithm

Data with similar pattern can be placed into same group through a process called clustering, which is an unsupervised learning algorithm. Clustering can partition unlabeled data into similar groups. Analysis of data and retrieval of information is the core task of cluster analysis (Xu and Wunsch 2005). The application of clustering algorithm in finance includes market segmentation, prediction of bankruptcy and scoring of credit. Hence, the clustering process can be applied extensively for the splitting of a large database into multiple clusters to discover the interesting pattern.

K-means, a simple unsupervised algorithm, capable of addressing the clustering problem (Wu et al. 2014). K-means is a partition algorithm which can handle a large database with multiple objects and this algorithm can generate the optimal cluster quicker. The algorithm runs simply by way of dataset classification and split the data into a number of fixed clusters (e.g., k cluster) in advance. Identifying the k centers is the basic operation of this algorithm which needs one for every cluster and the centers are located in a complicated way so that different result is produced by dissimilar location.

Therefore, the clusters need to be placed in a more distant location for better acceptable result. The next phase of k-means joins the nearest center of the dataset for every point. However, early grouping is formed to finish the first phase if no underlying point is available. Next, the re-calculation of k new centroids is required as barycenter for the available clusters from previous phase. At this stage, k new centroids are available and a new binding is required between nearest center and data points underlying same dataset to produce a loop. The loop facilitates the alteration of the location of k centers step by step so that no further alterations are possible.

The collected dataset from the stock market consists multiple organizations that needs to select one organization and k-means clustering algorithm is applied to the preprocessed dataset to select one organization. The observation of dataset produced through k-means clustering categorizes the organizations into fast growing and slow growing.

10.3.3 Input Features with Stock Data

Supervised learning method is chosen in this research, where the model is trained through a target attribute known as output. The output or target attribute is chosen as Investment Decision for all stock market, High price for NYSE, NASDAQ and Bursa Malaysia and Average price for DSE. The attributes collected from stock market and other factors for stock prediction are described as below:

Stock Number: The number, which is provided to a company during the enlistment with the stock market, is known as Stock Number. For an instance, British American Tobacco (BAT) receives stock number 4162 in Bursa Malaysia.

Stock Name: The name by which investor can recognize a company and also the name provided during enlistment is known as Stock Name. For an instance, British American Tobacco Malaysia is recognized as 'British American Tobacco' in Bursa Malaysia.

Date: The trading day when the stock market performs the trading operation is known as Date. For an instance, NYSE is open on January 2, 2018 at 9.30 a.m. local time.

Open Price: A security is first traded on a price for a particular trading day immediately after the opening of the stock market, which is known as Open Price. For an instance, NYSE opens at 9.30 a.m. local time and each security is traded at that time on an open price. Daily opening price is the first trade price for a listed stock.

Closing Price: The closing price represents the final trading price of a security for a particular trading day. It denotes the most recent valuation of a security till the commencement of next trading day for a stock market.

High Price: High Price represents the highest trading price of a security for a particular trading day. Usually, High Price is higher than the open or closing price of that security in a stock market.

Low Price: The lowest trading price of a security at a given trading day is identified as Low Price. Generally, Low Price is lower than the open, closing and high price of that particular security.

Average Price: The average price of Open, Close, High and Low price of a security is known as Average Price for a security.

Volume: The number of shares traded for a security or whole stock market during certain period of time is known as Volume. In stock trading, there is a seller for every buyer and total volume is calculated through number of transactions. If sellers and buyers agree for a transaction at an agreed price for certain number of shares, it makes one transaction and volume is determined through the number of transactions multiplied by number of shares.

Number of Trade: The number of trade or transaction took place for a security on whole stock market during certain period of time is known as Number of Trade.

Turn Over: The amount being traded for a security on a whole stock market during certain period of time is known as Turn Over.

Gold Price: The price of one gram of gold for a particular day is considered as Gold Price. In this research, gold price is measured through local currency based on same stock market.

Dollar Price: The exchange price of a local currency against US Dollar (USD) is known as Dollar Price. In international market, different foreign currencies are traded in terms of number of units per USD.

Bank Interest Rate: The amount paid to deposit holders by bank or financial institutions are known as Bank Interest Rate. The Bank Interest Rate is expressed as percentage of principal on annual basis.

FDI: The amount of investment made for establishing business or acquiring business assets by an individual or a company of one country to another country is known as Foreign Direct Investment (FDI).

Inflation: The rate at which the price of goods and services is rising and the purchasing power of currency is falling is known as Inflation. Usually, central bank keeps track of Inflation rate to limit inflation and avoid deflation so that the economy runs smoothly.

10.3.4 Target or Output in Stock Data

In building predictive model, selection of target variable is one of the preliminary steps which is a simple and straightforward process. In this research, the target or output variables are high price or average price and the prediction of price is made for a day ahead as per the recommendation made by Xing et al. (2017). The decision (buy, sell or hold) is provided to the investor as an additional information to determine whether a stock is suitable for him or her to make investment. In predicting the high price or average price, the closer the value availed through prediction is the better for investor. The decision whether to buy, sell or hold the stock, is calculated through the consultation with expert in stock field and also through the observation of the stock movement.

The output high price or average price and stock investment decision obtained through the experimentation are included in result section. The data is arranged in such a way that the actual output is separated for high price or average price prediction while investment decision is placed at last. The resulting output from the neural network model is the predicted output for this research.

10.3.5 Data Partitioning

The successful predictive models can be built through feed-forward neural network, which is a powerful neural network structure proficient in modeling prediction class from a nonlinear predictor attributes combination. The over-fitting problem needs to be tackled though the network is able to fit accurate model from normalized data. The over-fitting is a situation when the network does not have capability to generalize between input–output patterns (Haykin et al. 2009).

Before the feed-forward neural network was trained, the data used for training, testing and validating the network was divided using the divider and commands of MATLAB illustrated through the Function (10.2). The command codes cycle samples between the training set, validation set, and test set according to percentages. Where, training set is used to determine the optimal set of connection weights, test set is used to determine the appropriate network configuration and validation set is needed to measure the generalization capability of the model.

Maier and Dandy (2000) investigated through the review of previous researches that data can be divided in any percentage without considering statistical properties. However, it is difficult to evaluate the optimum result. Hence, the dataset is distributed initially 70% of the samples to the training set, 15% to the validation set and 15% to test set because the validation and test set requires same percentage (Maier and Dandy 2000).

$$\begin{aligned}&[\text{trainInput\_Data, valInput\_Data, testInput\_Data, trainInd, valInd, testInd}]\\&\quad = \text{dividerand(Input\_Data)}\end{aligned} \tag{10.2}$$

10.3.6 Evaluation of the Predicted Stock Price

The performance measurement of the proposed model is calculated applying Mean Absolute Percentage Error (MAPE), Mean Absolute Deviation (MAD) and Root Mean Squared Error (RMSE). Equations (10.3), (10.4), and (10.5) are used to calculate the MAPE, MAD and RMSE. Where, the actual values of the stock are (a_1, a_2, a_3, ..., a_n) and the predicted value of the stock are (p_1, p_2, p_3, ..., p_n).

$$\text{MAPE} = 100 * \frac{1}{n}\sum_{i-1}^{n}\left|\frac{a_i - p_i}{a_i}\right| \tag{10.3}$$

$$\text{MAD} = \frac{\sum_{i=1}^{n} |a_i - p_i|}{n} \tag{10.4}$$

$$\text{RMSE} = \sqrt{\text{mean}(a_i - p_i)^2} \tag{10.5}$$

The MAE closest to zero indicates that MAE is the better for ensemble model or else the closest the predictions to the actual value.

10.3.7 Implementation of MGWO Algorithm for Feature Selection

All the attributes from the collected dataset may not be significant for prediction and hence attributes selection is crucial in stock prediction research. Stock price may be influenced by numerous factors and predictors may encounter difficulties in selecting the input for experimentation. Witten et al. (2016) have suggested to select the attributes based on deep understanding of the learning problems at hand and real meaning of the available attributes (Witten et al. 2016).

Numerous methods are available for feature selection, whereas meta-heuristic algorithm can be a better option for feature selection which is also supported by Emary et al. (2016). In this research, the approach for MGWO are employed in the feature selection domain for finding feature subset maximizing the classification accuracy while minimizing the number of selected features. Wrapper's approach for feature subset selection suggests three main processes (Emary et al. 2016):

1. Classification method.
2. Feature evaluation criteria.
3. Search method.

In order to represent the population of each particular entry of MGWO inspected data, decision variables are defined to represent individuals, which are neural network parameters and inserted feature. This process is performed during each iteration of analysis. In other words, the algorithm inspects one feature in every iteration and provides an index value to rank that feature for further extraction process. Steps of MGWO implementation for feature selection are as indicated in Fig. 10.2.

Equation (10.6) presents the cost value, cost of seed parameters of neural network used in each classification cycle ω as well as the number of inserted features (which is in our case 16 features). During inspection of each feature, the feature k will be checked against the classification rate F_k as a tagged value ≥ 0.5 which will determine how accurate is the classifier given the selected feature set, then the feature will be selected in indicating selected feature k, otherwise it will be excluded from the list of final features.

$$z : F_k = [\text{Cost}\, \omega f_1 f_2 \ldots f_n] \tag{10.6}$$

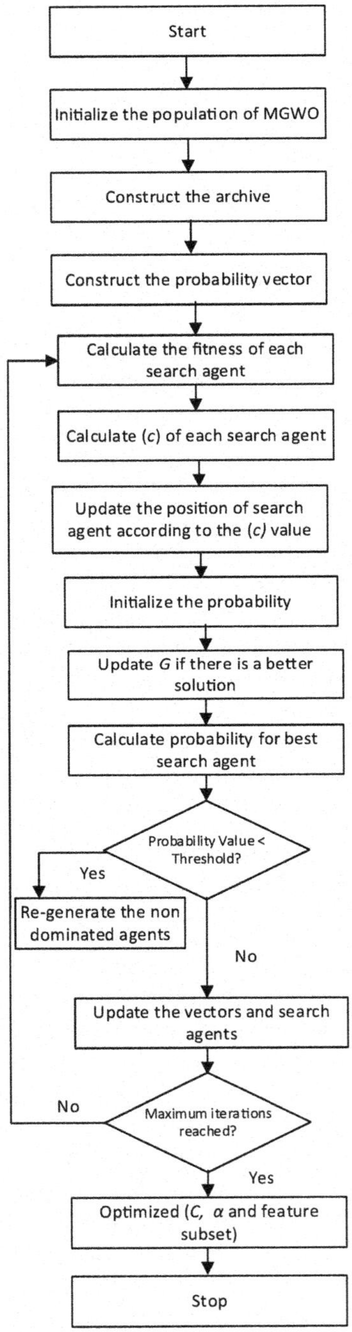

Fig. 10.2 Steps of MGWO implementation for feature selection

Each search agent calculates its fitness value upon selecting a subset of features F_k and gets compared against the index threshold value during training and testing processes. By applying this process during all iterations, eventually less competent features would be extracted as they have produced less impact on the obtained fitness value while the dominantly high indexed features would be kept in the final extracted list.

In this research, the selected features are confirmed through the consultation with three domain experts, who are the manager in securities division of bank related to DSE. Domain experts can guide well in selecting appropriate attributes for prediction (Suh 2012) and the information collected through the expert views for selecting the attributes of stock prediction are really useful in this regard.

10.3.8 Neural Network Model Design

In the current research, feed-forward multilayer perceptron architecture as illustrated in Fig. 10.3 is used for designing the neural network model. The data is divided as training, validation and testing as explained in Sect. 10.3.5. MGWO algorithm is used for training neural network to optimize MLP parameters.

The Mean Absolute Percentage Error (MAPE), Mean Absolute Deviation (MAD) and Root Mean Squared Error (RMSE) are used for performance evaluations which are objective function as well. Error is computed during each iteration of training so that error can be tracked easily and rising of error will terminate the iteration. Increasing of error indicates that the training process has converged.

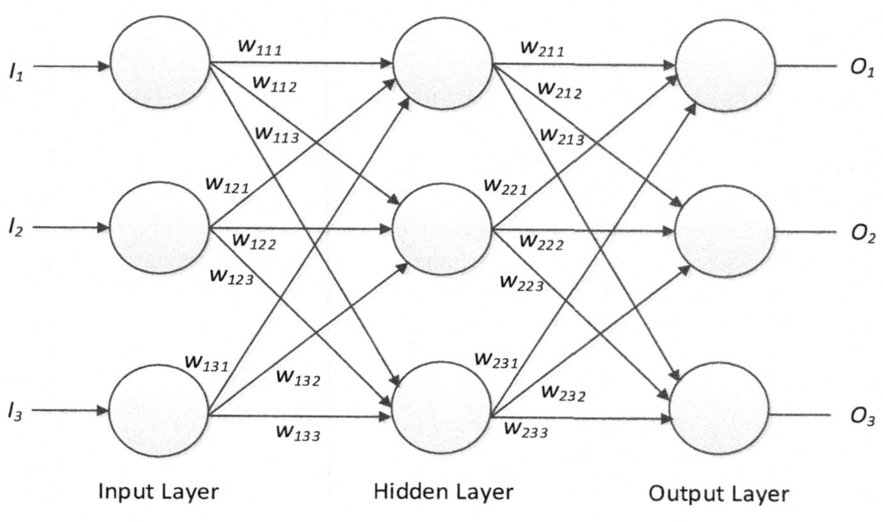

Fig. 10.3 Multi-layer perceptron architecture

In Fig. 10.3, the input layer is represented by I_1, I_2 and I_3 inputs, next the hidden layer which is not visible to users and it contains number of nodes, accepts value from preceding layer, executes mathematical operation on those values and pass it to the next layer which is the output layer that produces O_1, O_2, and O_3 outputs. In this architecture, each node calculates the sum of the value received from preceding nodes, performs the validation checking against a threshold value and produces the outputs by multiplying with layer weights W_{111}, W_{112}, W_{113}, W_{121}, W_{122}, W_{123}, W_{131}, W_{132}, W_{133}, W_{211}, W_{212}, W_{213}, W_{221}, W_{222}, W_{223}, W_{231}, W_{232}, W_{233} iteratively till the network converges and eventually the error starts rising at this point.

The activation function or transfer function represents the output where a function that increases the values to balance linear and nonlinear nature called as sigmoid function which is used for the whole network. If θ_m is the activation function for the output of a neuron in terms of the induced local field a and s is the slope of the function then the sigmoid function is represented by the following Eq. (10.7).

$$\theta_m = \frac{1}{1 + \exp(-sa)} \tag{10.7}$$

To avoid the over-fitting problem in this research, numbers of neurons are limited to 10 and the error is set to 0. Whereas, the maximum fails is fixed to 10 which facilitates the network to converge, if unable to meet up with other settings after a trial of 10 times. The iteration is set to 200, though the network may converge within a few numbers of epochs, if the configuration settings are fulfilled. The performance of neural network model is improved by training it using MGWO algorithm.

10.3.9 Ensemble of MGWO and Neural Network Algorithm

Figure 10.4 illustrates the ensemble of MGWO algorithm and neural network for feature selection and training Multi-Layer Perceptron neural network. The selected feature is forwarded to the MLP neural network that will be used to produce the best trading result through the learning of MGWO. Here, the predicted trend is evaluated to measure the performance of the prediction. In ensemble algorithm, the MGWO optimization process helps to determine the best feature combinations as indicated in Sect. 10.3.5 and then facilitates the MLP neural network to select the features so that the network can determine the best set of features for prediction.

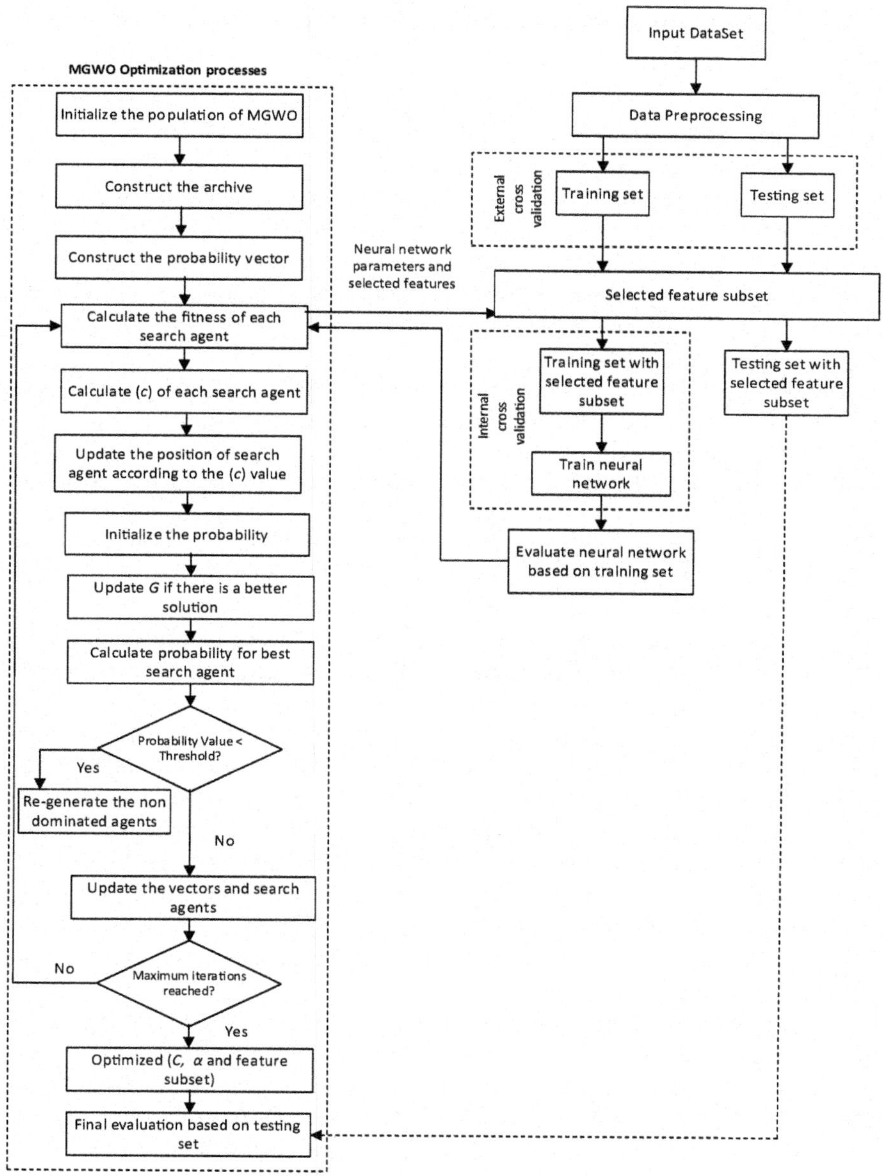

Fig. 10.4 Classification process through MGWO

10.3.10 Implementation of Ensemble Algorithm for Stock Prediction

In ensemble of neural network and MGWO, the first phase involves the selection and pre-processing of the datasets. It includes conversion of the stock data into continuous series and selection of the data for related factors such as bank interest rate, FDI, and gold rate. Besides, data from different information sources is converged into one substantial dataset, so all required data is accessible for performing simulation.

A k-means cluster data mining based decision support model is included so that the companies of the stock market can be categorized into two ways such as: high growth and low growth.

The classification of stock data is performed applying MGWO to verify how well they are compatible for prediction, select features and learning. The learning algorithm is determined at this stage.

The prediction of the stock price using Nonlinear Autoregressive Exogenous neural network algorithm is performed that includes the creation of a neural network based decision support model so that the predicted stock price can be availed for both high growth and low growth organization. Finally, the predicted stock price is compared to evaluate the performance of the prediction.

After passing through the several steps mentioned above a conclusion regarding the proposed research questions would be possible to make. In particular, assessment of whether ensemble of neural network and MGWO model is suitable decision support model in the domain of stock prediction or not. Figure 10.5 illustrates the concept of ensemble of neural network and MGWO, where each technique complements one another to contribute for acceptable stock prediction model.

The research applies the ensemble model consisting of neural network and MGWO to alleviate the limitation of each technique. The limitation of neural network includes difficulty in training (Srivastava et al. 2015), entrapment in local minima, result accuracy and convergence rate (Mirjalili et al. 2014) while standard Grey Wolf Optimizer algorithm has drawbacks of low solving precision, slow convergence, and bad local searching ability (Yang et al. 2017).

The proposed ensemble algorithm is explained in Fig. 10.6, which can be used for the stock prediction. The algorithm has 3 parts where Part (a) represents k-means clustering to determine the stock to be chosen for prediction and stock investment decision, i.e., buy, sell or hold, Part (b) represents feature selection, learning, and classification, and Part (c) represents, stock prediction applying MLP algorithm and the evaluation of prediction result. The input of the algorithm needs the pre-processing of the data and output is the predicted stock price, Buy/Sell/Hold Decision, effect of various factors on Stock price and the evaluation result.

The algorithm starts with applying k-means clustering to select the stock for prediction. Then, the algorithm will use the rule to make the decision of whether to Buy/Sell/Hold the stock. Next, MGWO algorithm will be applied for the training of neural network through the selection of the best features, determine the suitability of stock data for prediction and learning. Later, MLP neural network is applied to

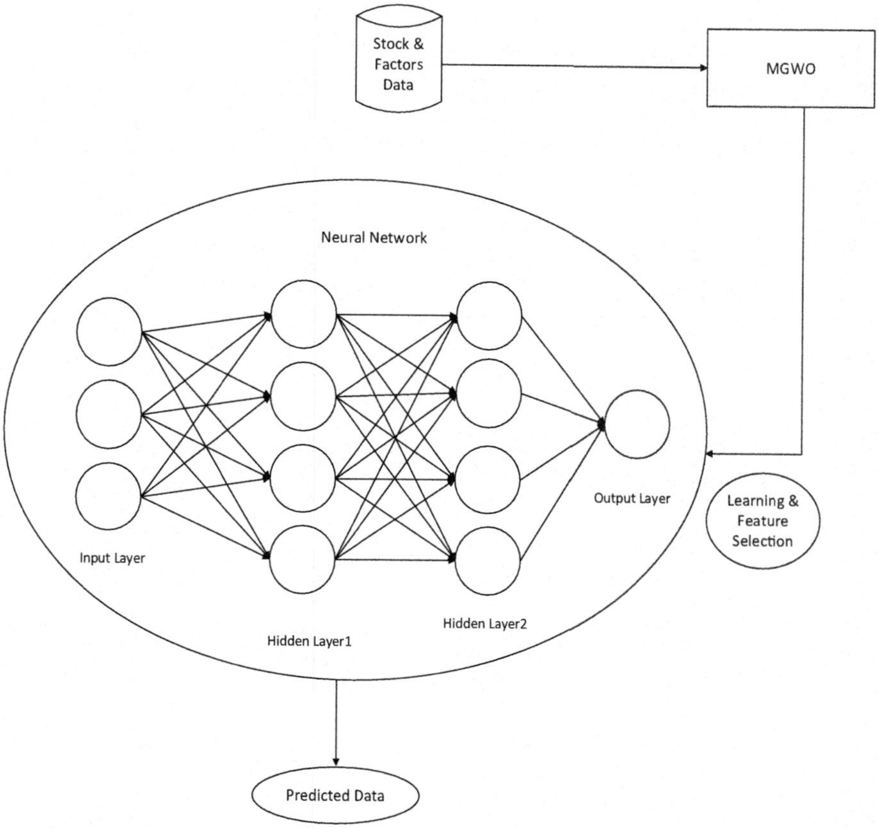

Fig. 10.5 Ensemble of neural network and MGWO for stock prediction

predict the stock price, various factors are integrated with stock data to determine the effect of various factors on stock price. Finally, the prediction result will be evaluated applying Eqs. (10.3), (10.4) and (10.5).

10.4 Computational Complexity of MGWO

The computational complexity of the MGWO depends on the number of generation (g), the population number (n), and the parameters dimensions (d). There-fore the overall computational complexity is O(MGWO) = O(Initialization) + g(O(Calculate the fitness of wolves) + O(Calculate the population in the archive) + O(Sort the population and archive population) + O(Select n best grey wolves from the population and archive population) + O(Update the population)). The computational complexity of initialization is $O(nd)$, the computational complexity of calculating the archive population and fitness is $O(n)$, the computational complexity of sorting the population and archive population is $O(2n \log 2n)$, the computational complexity of selecting n best grey wolves from the population and opposition population is $O(n)$, the computational complexity of updating the population is $O(nd)$. Therefore, the final computational complexity is $O(\text{MGWO}) = O(nd) + g(2O(n) + O(2n \log 2n) + O(nd))$.

```
**********K-means clustering to determine the stock to be chosen for prediction**********
Begin
Compute the distance of each data-point di (1<=i<=n) to all the centroids
cj (1<=j<=k) as d (di, cj)
 For each data-point di,
   Find the closest centroid cj and assign di to cluster j
 End For
Set ClusterId[i] = j        // j:Id of the closest cluster
Set Nearest_Dist[i] = d(di, cj)
For each cluster j (1<=j<=k), recalculate the centroids
  Repeat
End For
For each data-point di,
  Compute its distance from the centroid of the present nearest cluster
  If this distance is less than or equal to the present nearest distance, the data-point stays in the cluster
  Else
     For every centroid cj (1<=j<=k)
        Compute the distance d(di, cj);
     End For
     Assign the data-point di to the cluster with the nearest centroid cj
     Set ClusterId[i]=j
     Set Nearest_Dist[i]= d(di, cj)
End For
For each cluster j (1<=j<=k), recalculate the centroids
      Until the convergence criteria is met
End For
********** Determine the Buy/Sell/Hold Decision through the Stock Data **************
If Openprice2 < Openprice1, Highprice2<Highprice1, Lowprice2<Lowprice1,
    Closeprice2<Closeprice1 Then
      Decision = 2 (BUY)
Else If Openprice2 > Openprice1, Highprice2>Highprice1, Lowprice2>Lowprice1,
    Closeprice2>Closeprice1 Then
      Decision = 3 (SELL)
Else
      Decision = 1 (HOLD);
End If
Return (Decision)
```

(a)

Fig. 10.6 Proposed ensemble algorithm for stock prediction. **a** K-means clustering and determining buy/sell/hold decision. **b** Feature selection, classification, and learning through MGWO. **c** Predict the stock price, evaluate the factors, and evaluate the result

*********** *Feature Selection, Learning and Classification of the stock data through MGWO* ********

34: Initialize *p_size, max_iter, n, pos, flag*
35: Generate the *init_pos* of grey wolves randomly
36: Construct *A* archive of collected solutions in each iteration
37: Initialize **a, $\vec{A}$ and $\vec{C}$**
38: Initialize Controlling parameter of selecting/removing solutions from archive *A* (Selecting $G_\alpha{}^n P_1 = 1/N_i$ Removing $G_\alpha{}^n$ $P_2 = N_i$)
39: G_α = The grey wolf with the first highest fitness
40: G_β = The grey wolf with the second highest fitness
41: G_δ = The grey wolf with the third highest fitness
42: *Threshold=0.5*
43: $P(G_\alpha{}^n) = 1$, the probability vector, ($\Pi = \{G_\alpha{}^1, G_\alpha{}^2, G_\alpha{}^3, \ldots\ldots\ldots, G_\alpha{}^n\}$)
44: **While** *i<max_iter*
46: Calculate the fitness of grey wolves
47: **If** fitness $i_{th} < G_\alpha$
48: Update G_α with new fitness i_{th} value
49: Update G_α Position
50: **Else If** fitness $i_{th} > G_\alpha$ & fitness $i_{th} < G_\beta$
51: Update G_β with new fitness i_{th} value
52: Update G_β Position
53: **Else If** *fitness* $i_{th} > G_\alpha$ & fitness $i_{th} > G_\beta$ & $< G_\delta$
54: Update G_δ with new fitness i_{th} value
55: Update G_δ Position
56: **End If**
57: **For** $x = 1{:}p\_size$
58: **For** $y = 1{:}n$
59: **If** *pos(x,y) > 0.5*
60: *flag(y) = 1*
61: **Else**
62: *flag(y) = 0*
63: **End If**
64: **End For**
65: **End For**
66: Update the position of current grey wolf by, $\vec{G}(i+1) = (\vec{G}_1 + \vec{G}_2 + \vec{G}_3)/3$
67: Calculate the probability, $P(G_\alpha{}^n) = \frac{N(G_\alpha{}^n,\ A)}{\sum_{j=1}^{n} N(G_\alpha{}^j,\ A)}$
68: **If** $P(G_\alpha{}^n) < Threshold$
69: $v = abs(Max\ (G_\alpha) - \boldsymbol{fitness\_value})/\boldsymbol{fitness\_value}$ *%Normalization*
70: **For** $z = 1{:}\ p\_size$
71: Find non-dominated $P_1(G_\alpha) < v$
72: Update the archive *A*, remove $G_{\alpha_{th}}$
73: Re-Generate the *init_pos of* $G_{\alpha_{th}}$ randomly
74: **End For**
75: **Else**
76: Update *a*, **$\vec{A}$ and $\vec{C}$**
77: Calculate the fitness of grey wolves including selected features
78: Update G_α, G_β, and G_δ
79: **End If**
80: $i = i + 1$
81: **End While**
82: Return G_α, selected features

(b)

Fig. 10.6 (continued)

********* *Predict the Stock Price applying MLP neural network algorithm* ************
Partition the Stock Data,
    [A,P] = Stock_Data
    Anew = A(i : k)
    A = 1 : m
    P = 1 : m
Train the network using *MGWO*,
    net = narxnet(1:2,1:2,10)
[As,Ai,Pi,Ps] = preparets*(net,A,{},P);*
net = train(net,As,Ps,Ai,Pi);
view*(net)*
Calculate the network performance,
    [N,Af,Pf] = net(As,Ai,Pi)
perf = perform*(net,As,N)*
Run the prediction for necessary time steps ahead in closed loop mode,
    [netc,Aic,Pic] = closeloop(net,Af,Pf)
view*(netc)*
********** *Evaluate the Effect of Factors on Stock Price through Regression* **************
********* *The equation used for measuring the effect of factors on stock price,*

$D = p + qI$
Where, D = dependent variable, I = independent variable, q is the slope of the line and p is the M-intercept. **********************************

Calculate, $q = \frac{n*Sum(d*i)-Sum(i)*Sum(d)}{n*Sum(d^2)-(Sum(d))^2}$
Insert the values into the equation, $D = p + qI$

************************* *Evaluate the prediction result* ****************************

Calculate, $MAPE = 100 * \frac{1}{n} * (Sum(ABS(\frac{a_i - p_i}{a_i})))$
$MAD = \frac{Sum(ABS(a_i - p_i))}{n}$
$RMSE = SQR(Mean((a_i - p_i)^2))$
End

(c)

Fig. 10.6 (continued)

10.5 Chapter Summary

This chapter describes the methodology engaged in this book related to the research questions. In particular, the flow and implementation of Modified Grey Wolf Optimization algorithm. The chapter also explains the overview of stock prediction, process of stock categorization using k-means clustering data mining algorithm, classification using MGWO and ensemble of MLP neural network. The analysis showed that neural network is trained applying the MGWO for feature selection. The feature selection and training the neural network through the selection of features is illustrated with appropriate flowchart in this chapter. The MGWO is explained with the algorithm. The investigation involving ensemble algorithm of neural network and MGWO for stock prediction with the evaluation of the model is illustrated also in this chapter. Finally, the computational complexity of MGWO is discussed.

The next chapter will demonstrate the performance of ensemble model applying neural network and MGWO. The performance of ensemble model will be evaluated against the existing approaches along with the statistical analysis.

References

Atsalakis GS, Dimitrakakis EM, Zopounidis CD (2011) Elliott wave theory and neuro-fuzzy systems, in stock market prediction: the WASP system. Expert Syst Appl 38(8):9196–9206

Bonde G, Khaled R (2012) Stock price prediction using genetic algorithms and evolution strategies. Paper presented at the proceedings of the international conference on genetic and evolutionary methods (GEM)

Emary E, Zawbaa HM, Hassanien AE (2016) Binary grey wolf optimization approaches for feature selection. Neurocomputing 172:371–381

Gamil AA, Elfouly RS, Darwish NM (2007) Stock technical analysis using multi agent and fuzzy logic. Paper presented at the world congress on engineering

Hassan MR, Nath B (2005) Stock market forecasting using hidden Markov model: a new approach. Paper presented at the proceedings of the 5th international conference on intelligent systems design and applications

Haykin SS (2009) Neural networks and learning machines, vol 3. Pearson Upper Saddle River, NJ, USA

Kozdraj T (2009) Using artificial neural networks to predict stock prices. Folia Oecon 225:295–308

Kuo RJ, Chen C, Hwang Y (2001) An intelligent stock trading decision support system through integration of genetic algorithm based fuzzy neural network and artificial neural network. Fuzzy Sets Syst 118(1):21–45

Lahmiri S, Boukadoum M (2015) An ensemble system based on hybrid EGARCH-ANN with different distributional assumptions to predict S&P 500 intraday volatility. Fluct Noise Lett 14(01):1–10

Lertyingyod W, Benjamas N (2016) Stock price trend prediction using artificial neural network techniques case study: Thailand stock exchange. Paper presented at the 2016 computer science and engineering conference (ICSEC)

Lopez MI, Luna J, Romero C, Ventura S (2012) Classification via clustering for predicting final marks based on student participation in forums. Paper presented at the 5th international conference on educational data mining (EDM), Chania, Greece. International Educational Data Mining Society

Mahajan KS, Jamsandekar SS, Kulkarni R (2015) Portfolio investment model using neuro fuzzy system. Int J Comput Sci Inf Technol (IJCSIT) 6(2):1819–1823

Maier HR, Dandy GC (2000) Neural networks for the prediction and forecasting of water resources variables: a review of modelling issues and applications. Environ Model Softw 15(1):101–124

Merh N, Saxena V, Pardasani KR (2011) Next day stock market forecasting: an application of ANN and ARIMA. IUP J Appl Financ 17(1):70

Mirjalili S, Mirjalili SM, Lewis A (2014) Let a biogeography-based optimizer train your multi-layer perceptron. Inf Sci 269:188–209

Narayanan B, Govindarajan M (2015) Prediction of stock market using ensemble model. Int J Comput Appl 128(1):18–21

Navale G, Dudhwala N, Jadhav K, Gabda P, Vihangam BK (2016) Prediction of stock market using data mining and artificial intelligence. Int J Eng Sci 6(6):6539–6544

Patel J, Shah S, Thakkar P, Kotecha K (2015) Predicting stock and stock price index movement using trend deterministic data preparation and machine learning techniques. Expert Syst Appl 42(1):259–268

Srivastava RK, Greff K, Schmidhuber J (2015) Training very deep networks. Adv Neural Inf Process Syst 2377–2385
Suh S (2012) Practical applications of data mining. Jones & Bartlett Publishers, USA
Tilakaratne CD, Mammadov MA, Morris SA (2009) Modified neural network algorithms for predicting trading signals of stock market indices. Adv Decis Sci 2009(125308):1–22
Van den Berg J, Kaymak U, Van den Bergh W-M (2004) Financial markets analysis by using a probabilistic fuzzy modelling approach. Int J Approx Reason 35(3):291–305
Vincent O, Bamiro K (2013) Fluctuations in stock market prices: what went wrong, its implications to Nigerian economy? Int J Comput Appl 63(19):13–20
Wang J-J, Wang J-Z, Zhang Z-G, Guo S-P (2012) Stock index forecasting based on a hybrid model. Omega 40(6):758–766
Wei L-Y (2013) A hybrid model based on ANFIS and adaptive expectation genetic algorithm to forecast TAIEX. Econ Model 33:893–899
Witten IH, Frank E, Hall MA, Pal CJ (2016) Data mining: practical machine learning tools and techniques. Morgan Kaufmann, USA
Wong W-K, Manzur M, Chew B-K (2012) How rewarding is technical analysis? Evidence from Singapore stock market. Appl Financ Econ 13(7):543–551
Wu J, Coggeshall S (2012) Foundations of predictive analytics. CRC Press, USA
Wu K-P, Wu Y-P, Lee H-M (2014) Stock trend prediction by using K-means and apriori all algorithm for sequential chart pattern mining. J Inf Sci Eng 30(3):669–686
Xing FZ, Cambria E, Welsch RE (2017) Natural language based financial forecasting: a survey. Artif Intell Rev 1–25
Xu R, Wunsch D (2005) Survey of clustering algorithms. IEEE Trans Neural Netw 16(3):645–678
Yang B, Zhang X, Yu T, Shu H, Fang Z (2017) Grouped grey wolf optimizer for maximum power point tracking of doubly-fed induction generator based wind turbine. Energy Convers Manage 133:427–443

Chapter 11
An Extensive Comparison of Metaheuristics When Training Neural Networks for Stock Prediction

Abstract This chapter presents the experiments based on three relevant goals such as firstly, application of k-means clustering algorithm to produce two clusters and benchmarking of MGWO against other existing meta-heuristic algorithms, secondly, stock prediction through MLP neural network, and thirdly, findings are verified using statistical analysis. It demonstrates how feature selection is made through MGWO application. Besides, the benchmarking of the Result with GWO with the comparison of GWO and MGWO for available results of stock data classification is demonstrated here. The effect of stock price on various factors like Gold Price, Dollar Price, Bank Interest Rate, Foreign Direct Investment (FDI), and Inflation are presented as well. Last but not the least, the performance measurement of prediction and the comparison of proposed model in comparison with existing works are also performed here in this chapter.

Keywords Parameters for meta-heuristic algorithms · K-means clustering · Feature selection through MGWO · Benchmarking the result · Stock prediction · Multi-layer perceptron (MLP) neural network · Validity threats

11.1 Introduction

In the previous chapter, the ensemble model design and implementation were illustrated and elaborated. Specifically, the necessary adaptation of GWO to the stock prediction problem has been highlighted. Then, the tuning process for MGWO has been described, that is, by introducing the archive to evaluate which solution provides better probability to proceed further for training and regeneration. Finally, based on process design of MGWO, the implementation of MGWO has been elaborated.

This chapter explains and presents the results obtained at various stages of this research. The research is performed in three stages and hence the results obtained from all the stages are included here with detail explanation. This chapter precisely illustrates the clustering through the k-means, classification through the MGWO and

D. Das et al., *Optimization Algorithms in Machine Learning*, Engineering Optimization: Methods and Applications, https://doi.org/10.1007/978-981-96-3849-9_11

comparison of classification using other methods i.e. PSO, GA, ACO, ES, PBIL, BBO, and stock prediction through MLP neural network.

The validation of the result and the error associated with each model are also discussed in this chapter. Finally, the comparative benchmarking experiments with well-known strategies that are presented along with the necessary statistical analysis. The stock prediction approaches reinforced through literature review and the key findings related to the approach are demonstrated in detail here.

11.2 Experimentations

The experiments contain three relevant goals. Firstly, k-means clustering algorithm is applied to produce two clusters and benchmarking of MGWO against other existing meta-heuristic algorithms are performed. Secondly, stock prediction through MLP neural network is made. Finally, findings are verified using statistical analysis.

The comparisons are made as per the well-known benchmarks applied by prior researches (Emary et al. 2016, 2018; Eswaramoorthy et al. 2016; Cai et al. 2015; Delnavaz 2014; Mirjalili et al. 2014b).

The approaches described in this book has been implemented using the Weka Data Mining software and MATLAB. The results are presented in several tables and graphs. Tables 11.2, 11.3, 11.4, 11.5, 11.6, 11.7, 11.8, 11.9, 11.10, 11.11, 11.12, 11.13, 11.14 and 11.15 show the results obtained in the experiments. For fair comparison, Table 11.1 shows the parameters that are adopted for the existing meta-heuristic algorithms such as GA, ACO, PSO, PBIL, ES, BBO, GWO.

11.3 Result Through K-Means Clustering

In this research, k-means clustering algorithm is applied to produce two clusters. Based on Volume of Trades for a company, one cluster will contain fast growing companies and other will contain slow growing companies. The stock data used for experimentation is as per the sample shown in Table 11.1. Here, Tables 11.2, 11.3, and 11.4 demonstrate the k-means clustering outcome where, the stock with highest number of volume traded is placed in one cluster and the stock with average volume traded is placed in another cluster through k-means clustering algorithm.

It has been observed that ACI from DSE, Digi from Bursa Malaysia and Microsoft from NASDAQ have highest volume traded overall and hence those can be selected for stock prediction through Classification and MLP neural network as the investors have the higher possibility of gain through these stocks.

Table 11.1 Parameters for existing meta-heuristic algorithms

Algorithm	Parameter	Values
GA	Maximum number of generations	300
	Population size	1000
	Type	Real coded
	Selection	Roulette wheel
	Crossover	Single point (probability = 1)
	Mutation	Uniform (probability = 0.01)
ACO	Maximum number of iterations	300
	Population size	1000
	Initial pheromone (τ_0)	1e−06
	Pheromone update constant (Q)	20
	Pheromone constant (q_0)	1
	Global pheromone decay rate (p_g)	0.9
	Local pheromone decay rate (p_t)	0.5
	Pheromone sensitivity (α)	1
	Visibility sensitivity (β)	5
PSO	Maximum number of iterations	300
	Population size	1000
	Topology	Fully connected
	Cognitive constant (C_1)	1
	Social constant (C_2)	1
	Inertia constant (w)	0.3
ES	Maximum number of iterations	300
	Population size	1000
	λ	10
	δ	1
PBIL	Maximum number of iterations	300
	Population size	1000
	Learning rate	0.05
	Good population member	1
	Bad population member	0
	Elitism parameter	1
	Mutational probability	0.1
BBO	Maximum number of iterations	300
	Population size	1000
	Habitat modification probability	1
	Immigration probability bounds per gene	[0, 1]
	Step size for numerical integration of probabilities	1

(continued)

Table 11.1 (continued)

Algorithm	Parameter	Values
	Max immigration (I) and max emigration (E)	1
	Mutation probability	0.005
GWO	Maximum number of iterations	300
	Population size	1000
	$\vec{a}$	Linearly decreased from 2 to 0
	$\vec{A}$	Random values in $-2a$ to $2a$
	$\vec{C}$	Random values in 0–2

Table 11.2 K-means clustering output for DSE

Attribute	Full data (17,092)	Cluster # 0 (12,091)	Cluster # 1 (5001)
Company name	ACI	ACI	SQURPHARMA
LOWPRC	1309.27	280.95	3795.47
HIPRC	1348.31	293.25	3899.15
AVGPRC	1330.63	288.11	3851.13
CLSPRC	1330.63	287.91	3851.63
TRDVOL	110,681.94	147,207.60	22,373.25

Table 11.3 K-means clustering output for Bursa Malaysia (KLCI)

Attribute	Full data (7375)	Cluster # 0 (1475)	Cluster # 1 (5900)
Company name	BATM	BATM	Digi
LOWPRC	15.7035	57.4682	5.2624
HIPRC	15.843	58.0141	5.3002
OPNPRC	15.5668	56.9263	5.227
CLSPRC	15.7002	57.4475	5.2634
TRDVOL	6,646,949.6052	182,207.7458	8,263,135.07

Table 11.4 K-means clustering output for NASDAQ

Attribute	Full data (7608)	Cluster # 0 (5732)	Cluster # 1 (1876)
Company name	Microsoft	SPARTAN	Microsoft
LOWPRC	14.9738	6.7926	39.9707
HIPRC	15.3503	7.0333	40.7623
OPNPRC	15.1611	6.9147	40.3574
CLSPRC	15.1708	6.914	40.3989
TRDVOL	15,529,463.775	8,414,224.1626	37,269,630.8635

Table 11.5 Best set of features through MGWO

Instrument	Low price	High price	Open price	Closing price	Decision
IBM, NYSE	149.61	149.63	150.11	148.58	1 (Hold)
	150.26	149.25	151.95	149.22	1
	149.06	149.35	149.99	148.12	1
	149.07	148.25	149.6	148	1
	149.9	150.02	150.15	147.81	1
	151.43	150	151.6	149.65	1
	152.34	152.07	153.52	151.91	3 (Sell)
	150.51	152.52	152.96	150.25	1
	149.79	151.45	153.1	149.36	1
	149.95	148.41	150.41	148.32	1
	147.59	149.33	149.76	147.5	1
	147.75	148.4	148.65	147.23	1
	144.98	147.95	148.22	144.49	2 (Buy)
	149.61	149.63	150.11	148.58	1

11.4 Feature Selection Applying MGWO

As explained in Sect. 10.3.7, decreasing number of features to select key contributing features for better prediction is the main purpose of feature selection. In current research, MGWO with wrapper approach is applied for feature selection to provide better classification, faster convergence, and avoid overfitting. Consequently, number of features are reduced to 6 for prediction from total number of 16 selected features that have been gathered originally.

In the experimentation of feature selection, set of features have been provided for classification through MGWO initially and it produced different classification rate such as 80, 70, 60% and so on. As per the classification rate produced, a feature selection vector has been formed with the value 0–1 for each iteration. Finally, the set of features with best value has been chosen. This approach produced the best set of features as indicated in Table 11.5. Where, 1 represents 'Hold', 2 represents 'Buy' and 3 represents 'Sell'. The 'Decision' column with stock data has been incorporated by forming the rules in stock market investment.

11.5 Benchmarking the Result with GWO

This section demonstrates the comparison of GWO and MGWO with available results for stock data classification. Figure 11.1 demonstrates the performance analysis of the proposed MGWO benchmarked with other popular optimization algorithm. It

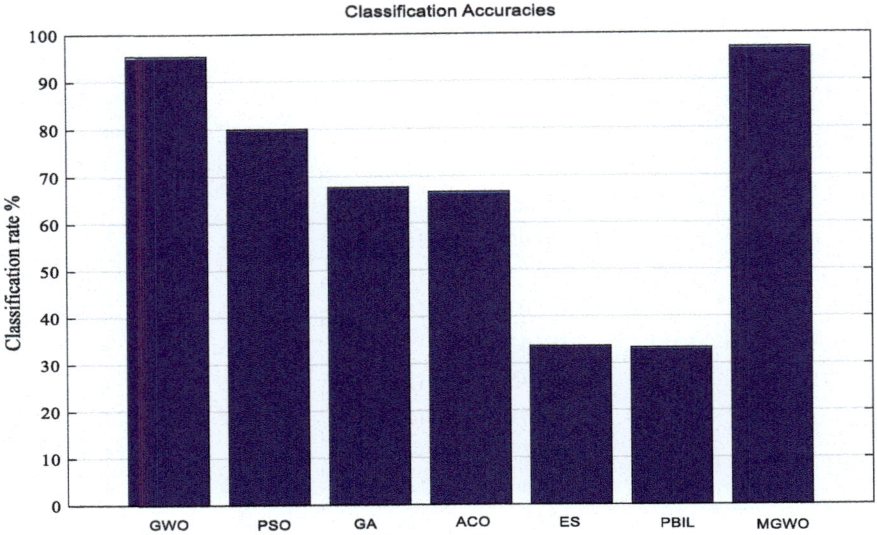

Fig. 11.1 Performance of MGWO in comparison with other popular optimization algorithms for stock market dataset

shows the classification accuracies along with comparison of classification with other algorithms i.e. PSO, GA, ACO, ES and PBIL. The classification rate for the stock data through MGWO is about 97% whereas, the classification rate through other algorithm is much lower. Moreover, the convergence graph shows that MGWO converge much faster than the compared meta-heuristic algorithms.

The reason for such an improvement of convergence is: firstly, due to the strengthening searching process by several random leaders in every iteration by MGWO whereas, GWO chooses the best alpha and beta, alpha follow. Secondly, due to the introducing of archive concept with the probabilistic model during the initialization phase that speeds up the convergence trends and enrich the quality of solution or accuracy. Thirdly, due to the re-generating random leaders in each iteration based on the statistical analysis performed on the collected fitness values in archive where, generation of wolf leaders will be highly randomized at the beginning of the hunt. This strategy can essentially improve the exploration power in the modified GWO from the early phase of iterations. Hence, the classification result through MGWO confirms that the selected stock data is suitable for applying neural network prediction algorithm to predict stock price (Table 11.6).

Table 11.6 Experimental result for the stock dataset

Algorithm	Classification rate (%)
GWO-MLP	95.3333
PSO-MLP	80
GA-MLP	67.6667
ACO-MLP	66.6667
ES-MLP	33.6667
PBIL-MLP	33.3333
MGWO-MLP	97

11.6 Stock Prediction Results Through MLP Neural Network

The six years of daily stock price for companies from NYSE, NASDAQ, Bursa Malaysia and DSE has been provided to predict the High Price through MLP neural network model which categorizes the data into Training, Validation and Test Set as indicated in Table 11.7. Where, Training Set consists 70% of total data, Validation Set contains 15% of data and Test Set holds 15% of data. The neural network model indicated in Fig. 11.2 contains 10 Hidden Neurons and d is Number of Delays which is 2 in this architecture.

MGWO is applied to train the network to fit the inputs and targets. The training produces the model as indicated in Fig. 11.3 which shows Training State (Plot train state), that Gradient = 0.39235, which is the calculation of weights used in network at epoch 28, which is one forward pass and one backward pass of all the training

Table 11.7 Categorization of dataset

Dataset type	Amount
Training set	70% of the target time steps
Validation	15% of the target time steps
Testing	15% of the target time steps

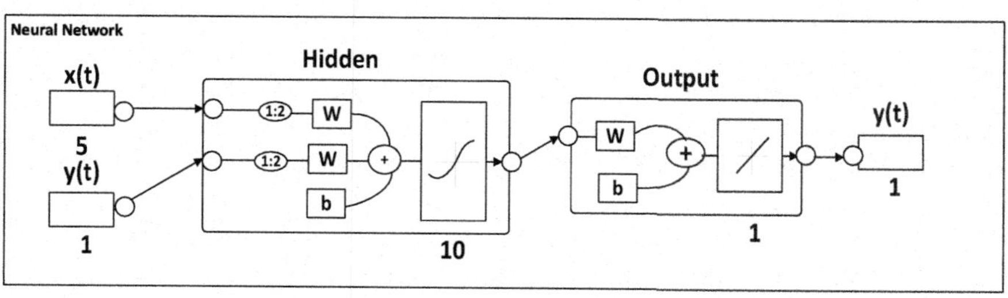

Fig. 11.2 Architecture of the neural network model

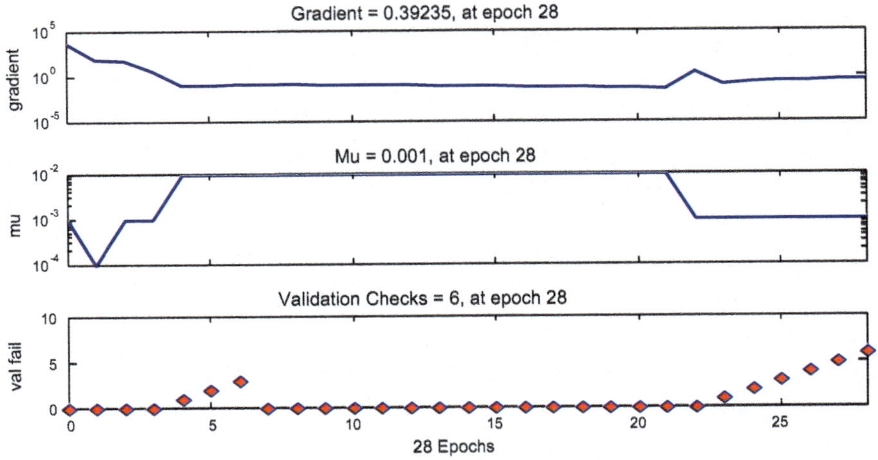

Fig. 11.3 Training state, gradient = 0.39235 at epoch 28. Mu = 0.001, at epoch 28. Validation checks = 6, at epoch 28

examples, Mu = 0.001 which is the control parameter for the algorithm used to train the neural network, at epoch 28. Validation checks = 6, at epoch 28.

The performance of the network is plotted in Fig. 11.4. For different combinations of data and parameters, this performance curve varies. Training of the model stops when it reaches to mentioned number of epochs (shown in Fig. 11.4) or alternatively, when Mean-Squared Error (MSE) is almost never improving after certain epochs. The circle in the performance curve shows the best validation performance.

Figure 11.5 represents the regression that facilitates defining nonlinear relationships in the experimental data and it can plot four regressions, demonstrating the network output with respect to actual data (target) for training, validation, test and all datasets. For a perfect fit, the data should fall along a 45° line, where the network outputs are equal to the targets (Mathworks 2012). Here, most data fall along 45° line and all the R values produced by each plot is more than 0.98 i.e. Regression by Training is 0.99273, by Validation 0.99356, by Test 0.98806 and by All 0.99194. Hence, it indicates that the fit by Regression is reasonably good for all datasets.

The time series response is plotted in Fig. 11.6 which indicates that there is not much variation between training target, training outputs, validation targets, validation outputs, test targets, test outputs. We can observe that they tend to have similar patterns.

Table 11.8 shows Actual Price versus Predicted Price of Stock Prices through MLP neural network.

The Time Lag of prediction is formally stated as: we find a function, $P: R^d \rightarrow R$ such as to obtain an estimate of $P(k)$ as indicated in Eq. (10.6).

The final predicted closing price of ACI generated through MLP neural network is listed in Table 11.9, which demonstrates Actual Price versus Predicted Price.

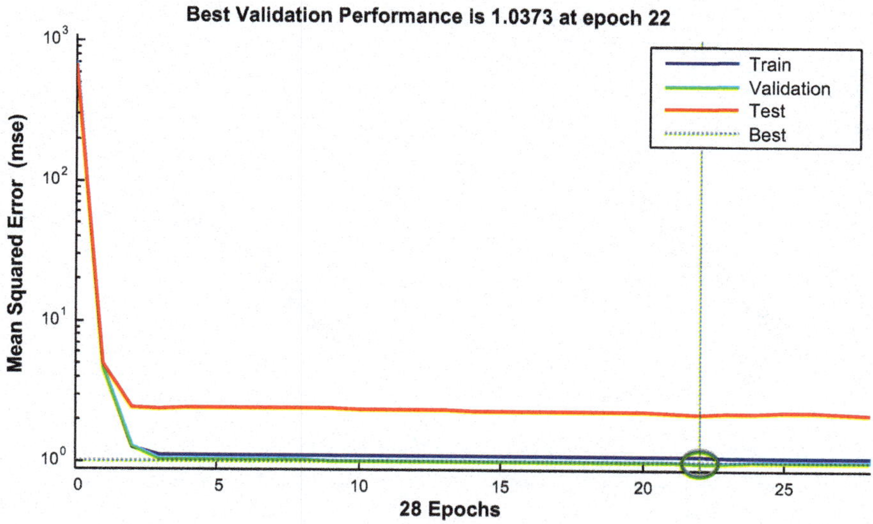

Fig. 11.4 Validation performance of the network

Figure 11.7 demonstrates the price of ACI for 6 years (A day in October). Table 11.10 illustrates Actual Price versus Predicted Price of stock price through MLP neural network for different stock market for a day, where the stock value for NYSE and NASDAQ in USD, stock value for Bursa Malaysia in Ringgit Malaysia (RM) and stock value for DSE in Bangladesh Taka (BDT).

11.7 Effect of Various Factors on Stock Price

The effect of stock price on various factors like Gold Price, Dollar Price, Bank Interest Rate, Foreign Direct Investment (FDI), and Inflation has been measured and it has produced the output indicated in Table 11.11. The *P*-value 0.5 or higher specifies that the effect is stronger, otherwise the effect is not very strong. Here, the *P*-value produced through the experiment indicates that there are some effect of those factors in fluctuating the stock price. However, the effect is not so strong and hence the factors may not heavily effect in changing (increasing or decreasing) the stock price.

11.8 Performance Measurement of Prediction

The performance measurement of the neural network model using Mean Absolute Percentage Error (MAPE), Mean Absolute Deviation (MAD) and Root Mean Squared Error (RMSE) is calculated. Suppose (a_1, a_2, a_3, …, a_n) are actual values

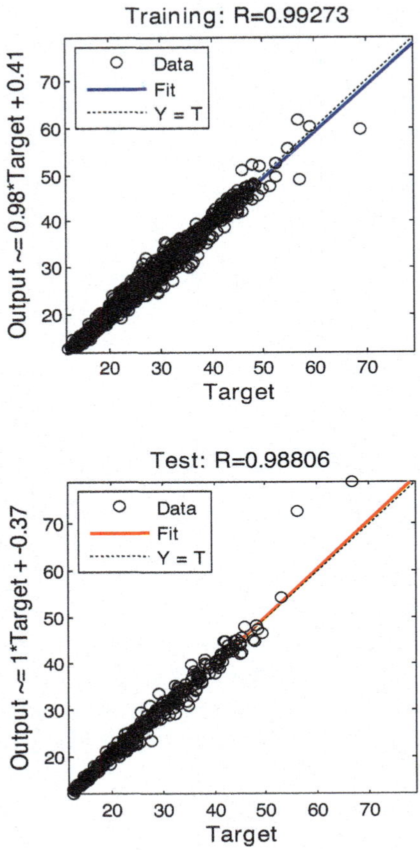

Fig. 11.5 Neural network output with respect to target (actual data) through regression

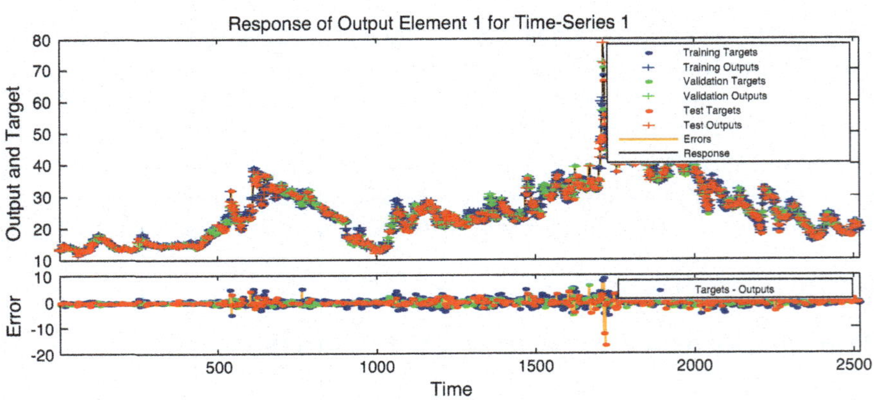

Fig. 11.6 Response of output element for time series

Table 11.8 Actual price versus predicted price of INTECH

Company name	Actual	Predicted	Prediction error (%)
INTECH	15	13.725	8.5
INTECH	14	14.0309	0.2
INTECH	14	13.6375	2.6
INTECH	13.7	13.9958	2.2
INTECH	13.9	13.8036	0.7
INTECH	13.8	13.4366	2.6
INTECH	13.5	13.6769	1.3
INTECH	13.6	13.6421	0.3
INTECH	13.6	13.6421	0.3
INTECH	13.6	13.9009	2.2
INTECH	13.8	13.9647	1.2
INTECH	13.9	14.1863	2.1

Table 11.9 Actual price versus predicted price of ACI

Company name	Date	Actual	Predicted	Prediction error (%)
ACI	2015-10-01	583.30	583.74	0.1
ACI	2015-10-04	573.10	575.77	0.5
ACI	2015-10-05	577.40	575.91	0.3
ACI	2015-10-06	575.20	574.01	0.2
ACI	2015-10-07	570.80	571.10	0.1
ACI	2015-10-08	568.50	570.02	0.3
ACI	2015-10-11	560.40	562.39	0.4
ACI	2015-10-12	562.50	564.48	0.4
ACI	2015-10-13	556.70	559.11	0.4
ACI	2015-10-14	554.50	556.42	0.3
ACI	2015-10-15	553.70	557.81	0.7

and $(p_1, p_2, p_3, \ldots, p_n)$ are the predicted values then the MAPE, MAD, and RMSE can be calculated using the equation indicated in (10.3), (10.4), and (10.5).

The evaluation of the neural network performance through the above equation is shown in Table 11.12.

In this work, INTECH Company's real price data is anticipated which was high level in Dhaka Stock Exchange. The stock data for INTECH is used for prediction through the neural network. After the network is created, the evaluation demonstrates a positive performance improvement, which is very encouraging for this research work and it will guide the investor towards investment in a particular security.

Table 11.13 demonstrates the performance evaluation for ACI that indicates positive performance improvement through the created network, which is encouraging

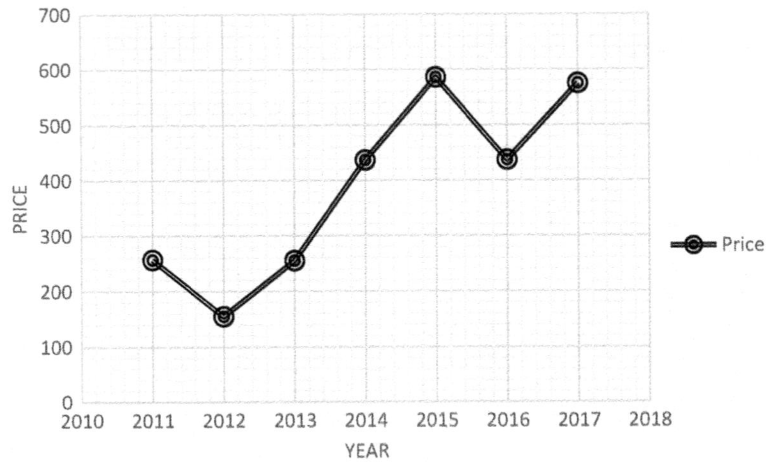

Fig. 11.7 Share price for ACI for 6 years (a day in October)

Table 11.10 Actual price versus predicted price for various stock markets worldwide

Company name	Stock market	Date	Actual	Predicted	Prediction error (%)
Microsoft	NASDAQ, USA	2016-12-01	60.15	59.71	0.72
IBM	NYSE, USA	2016-12-01	162.2	160.966	0.76
Digi	Bursa Malaysia	2016-12-01	4.96	4.986	0.536
ACI	DSE, Bangladesh	2016-12-01	414	415.37	0.33

for this research work as well for guiding the investor for investment into a particular stock.

The performance of the predicted price is evaluated through Eq. (9.5) for various instruments of different stock markets produced the Root Mean Square Error (RMSE) value as indicated in Table 11.14 and the Prediction error related to the prediction is indicated in Fig. 11.8. The RMSE value close to 0 indicates no error and prediction is completely acceptable. The Prediction error remains lower and reasonable through the proposed model. However, the prediction error for NASDAQ is little higher than other stock market which is 0.7624 due to the effect of factors. Hence, it can be concluded that the proposed ensemble model can be applied for stock prediction that can also reduce the error.

The performance of a neural network can be also evaluated through generalization capability (Kaastra and Boyd 1996). The proposed prediction model can perform well for various stock markets and hence it has good generalization capacity.

Table 11.11 Effect of various factors on stock price

Regression statistics								
Multiple *R*	0.0702883							
R square	0.0049405							
Adjusted *R* square	– 0.0022389							
Standard error	0.5830351							
Observations	699							
ANOVA								
	df	SS	MS	*F*	Significance *F*			
Regression	5	1.16960749	0.23392	0.68815	0.63254			
Residual	693	235.571451	0.33993					
Total	698	236.741059						
	Coefficients	Standard error	*t* stat.	*P*-value	Lower 95%	Upper 95%	Lower 95.0%	Upper 95.0%
Intercept	1.480864	0.03804066	38.9284	1E–176	1.40618	1.55555	1.40618	1.55555
Gold price	– 0.0388271	0.06284375	– 0.6178	0.53689	– 0.1622	0.08456	– 0.1622	0.08456
Dollar price	0.0210943	0.03016187	0.69937	0.48456	– 0.0381	0.08031	– 0.0381	0.08031
Bank interest rate	– 0.1545621	0.33800539	– 0.4573	0.64762	– 0.8182	0.50908	– 0.8182	0.50908
FDI	– 0.5084582	0.34658954	– 1.467	0.14282	– 1.1889	0.17203	– 1.1889	0.17203
Inflation	0.0380297	0.10948312	0.34736	0.72843	– 0.1769	0.25299	– 0.1769	0.25299

Table 11.12 Evaluation of prediction for INTECH

Input parameters (previous 10 average prices)	Neural network architecture	Forecasting performance	INTECH
P_{t-1}	MLP	MAPE	2.0
$P_{t-2}, \ldots, P_{t-9}$	MLP	MAD	0.3
P_{t-10}	MLP	RMSE	0.4

Table 11.13 Performance evaluation of ensemble model for ACI

Input parameters (previous 10 closing prices)	Neural network architecture	Forecasting performance	ACI
P_{t-1}	MLP	MAPE	0.28
$P_{t-2}, \ldots, P_{t-9}$	MLP	MAD	1.18
P_{t-10}	MLP	RMSE	1.75

Table 11.14 RMSE value for instrument of various stock markets

Company name	Stock market	RMSE
Microsoft	NASDAQ, USA	0.7624
IBM	NYSE, USA	0.0199
Digi	Bursa Malaysia	0.0645
ACI	DSE, Bangladesh	0.2249

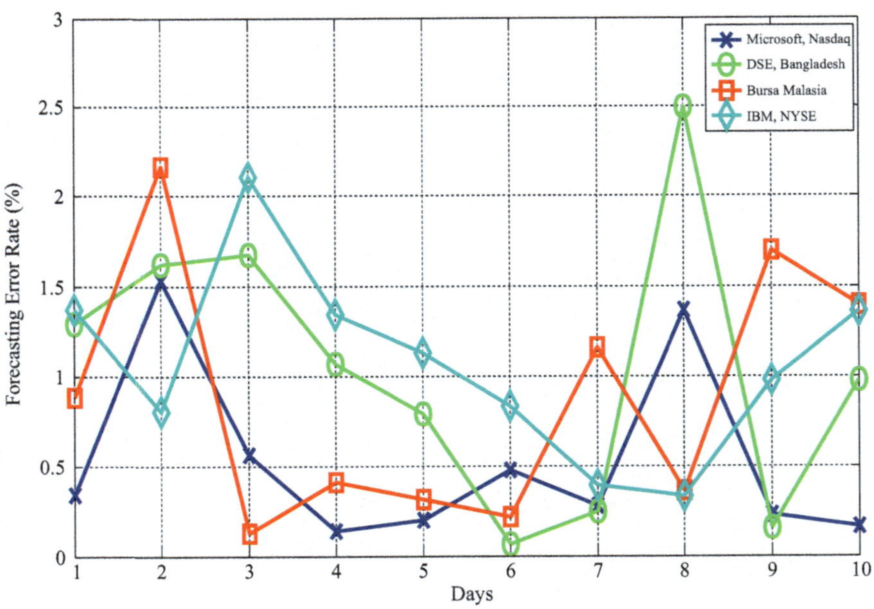

Fig. 11.8 Comparison of prediction error for various stock markets

11.9 Comparison of Proposed Model with Existing Works

The existing models tried to predict from the markets perspective by applying models such as technical analysis, fundamental analysis and linear regression; none of these however has proved to be consistent in making correct prediction with less error (Adebiyi et al. 2012; Lawrence 1997). These methods are based on base level standard. Technical analysis phase is very subjective and contradict the efficient market hypothesis (Lawrence 1997).

It is difficult to time the market through fundamental analysis and optimizing the approach is time consuming and hard to implement (Adebiyi et al. 2012; Lawrence 1997). In contrast, finding the global network minimum is not guaranteed through Back Propagation. Although error is not minimized through this, there is possibility of weight modification to meet local minimum in the error landscape, but the network may not be optimized (Lawrence 1997; Rojas 1996). Convergence through Back Propagation is very slow and not guaranteed. Moreover, learning requires input scaling and normalization (Budhani et al. 2012). Neural network approach is having a drawback of only 'learn' through past patterns which requires skilled tuning of the parameters. The basis of the stock price movements is very difficult to capture.

Over fitting is a serious problem (Haykin 1994) that occurs when the network has too many free parameters. These parameters allow the network to fit well with the training data but typically lead to poor generalization. There are two main reasons for this, first is due to having too many nodes to the networks and the second is due to the network being trained more. If the input data has high dimensions then NN is restrained in learning the patterns (Rojas 1996). Multi-layer perceptron has problems of getting stuck in a local minimum and it is very slow in learning (Mirjalili et al. 2014a).

In the proposed model, an ensemble model consisting of neural network and MGWO has been implemented. Here, k-means clustering is used to categorize the organization, classification to determine the suitability of data for prediction, feature selection, and learning. Then, MLP neural network algorithm is applied to predict the stock price. For stock prediction, neural network has been taken as preferred network instead hybrid model because it takes less time and performs well (Jia et al. 2013; Fasanghari et al. 2015).

After processing the result the error is calculated in percentage and then evaluated the performance. Fewer errors have been found in the case of the ensemble model that consists of neural network and MGWO when compared to other meta-heuristic models. Finally, the performance analysis phase demonstrates the accuracy of the model of the stock market.

According to the analysis, it has been realized that, more accurate matches amongst the data has been experienced through this research. Figure 11.9 shows the performance of the prediction through the proposed model, where the value of Actual, Predicted and Prediction Error have compared. It indicates that the prediction error is slightly high for first observation. However, the prediction errors are trivial for remaining observations and most of the Actual and Predicted values are much closer. A predictive model is just a guess of what will happen in future. Hence, experimental findings through a model will be acceptable if the error is less.

The Mean-Squared Error (MSE) value is also quite reasonable in proposed work i.e. 0.4. Provided that the lower MSE value is always better. MSE value 0 indicates that there is no error and the prediction is utterly acceptable (Yetis et al. 2014).

MGWO has produced good classification rate 97 which is better than other algorithms. Figure 11.10 demonstrates below the performance comparison between Proposed and Existing work in terms of prediction price. It indicates that the prediction is closer to actual price in proposed model applying MLP compared with existing models applying GWO, GA, ACO, PSO, PBIL, and ES (Navale et al. 2016; Hafezi et al. 2015; Billah et al. 2015; Khan et al. 2011). So, it can be concluded that, the ensemble of neural network and MGWO can reduce the errors and the prediction can be more accurate as well. Table 11.15 presents the comparison between existing and proposed research findings.

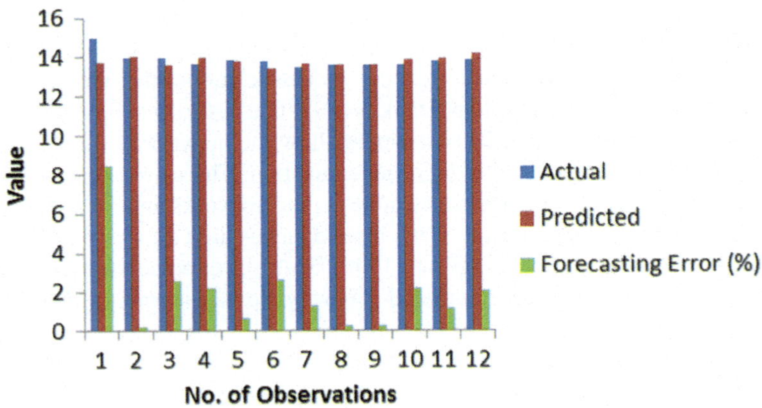

Fig. 11.9 Prediction performance

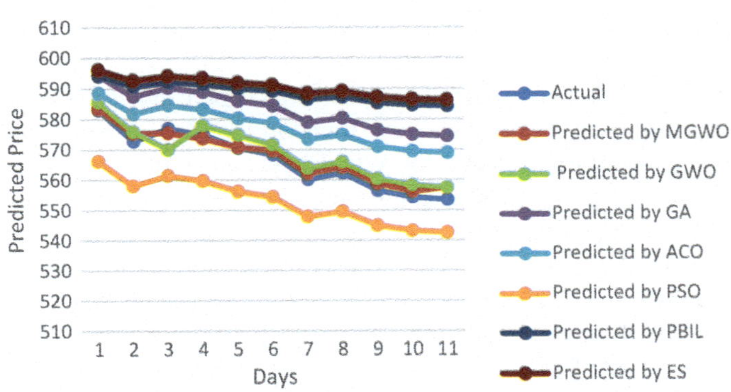

Fig. 11.10 Comparison of prediction performance

Table 11.15 Comparison between existing and current research findings

Research model	Year of research	Existing research finding	Current research finding
General stock prediction model	Xiong et al. (2014)	Statistical model i.e. multi-output support vector regression is used for stock index prediction (Xiong et al. 2014) which can deal with linear data but stock data is non-linear	Ensemble model can deal well with non-linear data
Stock prediction using data mining	Patel et al. (2015)	Data pre-processing and use of discrete data is emphasized for the improvement of prediction accuracy (Patel et al. 2015) where the algorithm gained about 50% accuracy whereas some algorithms remained silent for few situations i.e. when to sell the stock	Ensemble model consisting of neural network and MGWO is applied for stock prediction which can make about 95% accurate prediction
Stock prediction using neural network	Hafezi et al. (2015)	Bat Neural Network Multi Agent System (BNNMAS) is used to predict the stock price for DAX and gained significant result with good accuracy for long term period. However, the Mean Absolute Percentage Error (MAPE) remained 2.84 for such model (Hafezi et al. 2015)	Ensemble model can reduce the errors, avoids the problem of over-fitting or under-fitting and MAPE remained 2.0 or lower
Stock prediction using classification	Ballings et al. (2015)	Some studies have compared their result with buy-and-hold strategy and found the combination of technical analysis and classification produced more profits. Eventually, their proposed model also reduced the risk. However, they have suggested using of better classification algorithm such as SVM, and k-NN. to achieve more generalization (Ballings et al. 2015)	Ensemble model is a combination of neural network and MGWO which gains more accuracy in predicting and error is reduced as well. The model is applied to Asian market and compared with prominent market
Stock prediction using ensemble algorithms	Navale et al. (2016)	Some studies have availed about 77% accuracy in prediction applying combinatorial algorithms which is higher than the single use of ANN or DT. Researchers recommended to perform further research by applying other models such as SVM, ensemble of artificial intelligence and other similar algorithm to reveal the weaknesses of other researches to reveal the uncertainty of stock market (Navale et al. 2016)	The proposed ensemble model gained more accuracy which is about 95% through the combination of neural network and MGWO that provides more accuracy with reduced error rate

(continued)

Table 11.15 (continued)

Research model	Year of research	Existing research finding	Current research finding
Stock prediction covering various factors	Bonde and Khaled (2012), Negnevitsky (2005) and Hafezi et al. (2015)	Few studies have investigated the effect of some factors and indicated that various internal and external factors play important role in moving the stock price that is why stock prediction is so complicated	Current study combines the factors with stock data and investigates the effect of the factors on stock data to provide better guideline to investors
Generalization of stock prediction for various markets	Lertyingyod and Benjamas (2016) and Hafezi et al. (2015)	Most of the researches are concentrated on a single stock market. Lertyingyod and Benjamas (2016) investigated Thailand Stock Exchange and Hafezi et al. (2015) concentrated on DAX	The proposed research provides more generalization by predicting for various stock markets i.e. promising and emerging stock markets and making a comparison

11.10 Statistical Analysis for the Experimental Findings

To validate the findings through the experimentation, this section demonstrates the statistical analysis and comparisons for the obtained outcomes. The statistical difference between the existing and proposed approach is determined through two tests specifically, Friedman test and Wilcoxon signed-rank test. Precisely, Friedman test is implemented to distinguish contrasts between the findings through all approaches and Wilcoxon signed-rank test is applied based on the result of Friedman test for the analysis of each approach's significance.

Friedman test is conducted through two hypotheses namely, null hypothesis (H_0) and alternative hypothesis (H_1) as per the equation indicated in (11.1), (11.2) and (11.3).

$$H_0 : \mu_1 = \mu_2 = \mu_3 \ldots = \mu_n \tag{11.1}$$

$$H_1 : \mu_1 \neq \mu_2 \neq \mu_3 \ldots \neq \mu_n \tag{11.2}$$

$$x^2 = \frac{12c}{n(n+1)} \sum_{i=1}^{n} \left(R_i - \frac{(n+1)}{2} \right)^2, \tag{11.3}$$

where n is the number of approaches, μ_i is the median of results for ith approach, c is the number of comparisons, and R_i are the ranks approaches' results. In Eq. (11.3), if the Friedman test statistic x^2 > critical value, then H_0 will be rejected. Here, the critical value is calculated based on a probability threshold called Alpha (α) and Degree of freedom (df).

Friedman test has been conducted for the stock prediction samples in Table 11.9 for all the approaches namely, MGWO-based-ensemble, GWO-based-model, GA-based-model, ACO-based-model, PSO-based-model, PBIL-based-model and ES-based-model. The test produces the results presented in Table 11.16.

Table 11.16 illustrates the statistical analysis through Friedman test for stock prediction data. The analysis specifies that there are significant differences between exiting approaches and MGWO. For the experiment, null hypothesis (H_0) is rejected. Moreover, MGWO performs significantly better than GWO, GA, ACO, PSO, ES, and PBIL.

Wilcoxon Signed Rank Test analyses the significance of each approach. The test is conducted through two hypotheses namely, null hypothesis (H_0) and alternative hypothesis (H_1) as per equation indicated in (11.4) and (11.5).

$$H_0 : \mu 1 - \mu 2 = 0 \tag{11.4}$$

$$H_1 : \mu 1 - \mu 2 \neq 0, \tag{11.5}$$

where $\mu 1$, $\mu 2$ are the median for proposed approach and median for other approach respectively. H_0 in Eq. (11.4) implies that there is no significant difference between the two approaches', while H_1 in Eq. (11.5) specifies that there is a difference between the two approaches'. Here, the decision is made based on α or significance level.

Table 11.17 illustrates the statistical analysis through Wilcoxon signed-rank test for stock prediction data. For the experiment, null hypothesis (H_0) is rejected. The analysis specifies that there are significant differences between MGWO and existing approaches.

Table 11.16 Friedman test

Test statistics		Decision
Number of samples	11	Reject H_0 and there are differences between findings of each approaches
Degree of freedom (df)	7	
Critical value	14.06	
Chi-square (χ^2)	269.0000048	
Asymp. sig.	0.000	

Table 11.17 Wilcoxon signed rank test

Pairs	Ranks				Asymp. sig. (2-tailed)	Conclusion
	Negative	Positive	Ties	Total		
GWO-MGWO	66	0	0	66	0.9987	Reject H_0
PSO-MGWO	66	0	0	66	0.9987	Reject H_0
GA-MGWO	66	0	0	66	0.9987	Reject H_0
ACO-MGWO	66	0	0	66	0.9987	Reject H_0
ES-MGWO	66	0	0	66	0.9987	Reject H_0
PBIL-MGWO	66	0	0	66	0.9987	Reject H_0

11.11 Discussion

Developing a general model for stock prediction is very complex because of non-linear nature of stock data. There is no single predominant approach to perform stock prediction. Parameter tuning and feature selection can play vital role to achieve fair outcome. Application of meta-heuristic can be an effective approach in this field, however meta-heuristic algorithm needs to exploit the search operator to enhance the performance.

The choice of Grey Wolf Optimizer (GWO) as a basis of classification is worth mentioning here. The encouraging phase of GWO and MGWO are that the algorithms are formed based on simple concept consists of hunting preys by grey wolves in nature. Due to this, the application of the algorithm is straightforward. On the other hand, all meta-heuristic algorithms are not so easy to implement as an algorithm suits well with one may not perform well for other problems.

Moreover, choice of appropriate value for parameter is very challenging with meta-heuristic approach because the performance of the algorithm vigorously depend on parameter modification. Inappropriate parameter value may result expensive computational efforts with poor outcome. In addition, difficult approaches with numerous parameters require noteworthy endeavors for adjustment. For instance, GA needs adjustment of mutation rate, crossover rate, population size, but GWO and MGWO needs to adjust two parameters $\vec{A}$ and $\vec{C}$. Feature selection is another issue that needs attention as additional features may deteriorate the outcome. So, reduced and appropriate feature produce better prediction which is facilitated through MGWO.

Regarding the overall performance of MGWO and complete ensemble model for stock prediction, ensemble model with MGWO produces better result in comparison with other models. Additionally. MGWO can produce better result in comparison with GWO due to its exploration and exploitation capacity with archive, strengthen searching process by several random leaders and re-generating random leaders in each iteration. Indeed, MGWO can convergence well in compared to GWO. However, computationally MGWO may take longer time than GWO because of implementing archive.

Finally, the statistical test confirms that there are significant differences between MGWO and existing approaches. Besides, MGWO performs significantly better than GWO, GA, ACO, PSO, ES and PBIL.

11.12 Validity Threats

The research may accompany several validity threats with the experimental studies. Few threats have been detected in this study that may effect the results obtained through current research.

Firstly, the choice of benchmark is a crucial threat. The study implements the experimental benchmarks for various renowned researches undertaken earlier in literature. Although, the employed benchmarks are chosen from Bangladesh, Malaysia, NYSE and NASDAQ stock markets, the same technique proposed in this research can be applicable to other stock markets also.

Secondly, all the approaches GA, ACO, PSO, ES, PBIL, GWO, MGWO select 300 maximum number of generations and 1000 population size which can be a major threat to the experimentations due to unfair number of comparisons. Because, some of the approaches may complete the search earlier. We can eliminate this threat by not limiting the generations and population size, instead we can set the same maximum number of fitness function evaluation as a stopping criteria.

Thirdly, the meta-heuristic approaches implement random search operator which can be a threat as well. The optimum result can be determined just for once by chance. Hence, the comparison of the optimum result may not indicate the actual performance of an approach. The solution to this threat may be avoided by selecting the mean result instead of optimum result.

Another threat is the comparison with other approaches. Many approaches are adopted meta-heuristic algorithm for stock prediction. However, accommodating all the approaches is beyond the implementation of current study as MGWO is not benchmarked with all available approaches in literature. To overcome this threat, the current research selects the recently published journal to choose the approaches related to GWO and stock prediction.

Lastly, performance evaluation strategies can also indicate another threat. The internal structure of performance measurement algorithm differs for different approaches. However, the current study selects the recognized performance measurement in literature (i.e., for stock prediction).

11.13 Chapter Summary

This chapter demonstrated the experimental setup and evaluation of MGWO. Initially, MGWO is compared against meta-heuristic models such as PSO, GA, ACO, ES, PBIL and GWO. Then, the performance of a developed ensemble model is demonstrated through stock dataset. The stock prediction results through application of k-means clustering data mining algorithm, classification using meta-heuristic algorithm are also displayed. The results through the ensemble of MLP neural network and MGWO are also demonstrated and discussed thoroughly. The comparison of proposed research work with existing models and performance evaluation of the prediction are also illustrated in this chapter. Finally, the results from experiments are evaluated statistically applying statistical test model.

References

Adebiyi A, Ayo C, Adebiyi MO, Otokiti S (2012) Stock price prediction using neural network with hybridized market indicators. J Emerg Trends Comput Inf Sci 3(1):1–9

Ballings M, Van den Poel D, Hespeels N, Gryp R (2015) Evaluating multiple classifiers for stock price direction prediction. Expert Syst Appl 42(20):7046–7056

Billah M, Waheed S, Hanifa A (2015) Predicting closing stock price using artificial neural network and adaptive neuro fuzzy inference system ANFIS: the case of the Dhaka stock exchange. Int J Comput Appl 129(11):1–5

Bonde G, Khaled R (2012) Stock price prediction using genetic algorithms and evolution strategies. Paper presented at the proceedings of the international conference on genetic and evolutionary methods (GEM)

Budhani N, Jha C, Budhani SK (2012) Application of neural network in analysis of stock market prediction. IJCSET 3:61–68

Cai Q, Zhang D, Zheng W, Leung SC (2015) A new fuzzy time series forecasting model combined with ant colony optimization and auto-regression. Knowl-Based Syst 74:61–68

Delnavaz B (2014) Forecasting of the stock price index by using fuzzy-neural network and genetic algorithms. J Appl Sci Agric 9:109–117

Emary E, Zawbaa HM, Hassanien AE (2016) Binary grey wolf optimization approaches for feature selection. Neurocomputing 172:371–381

Emary E, Zawbaa HM, Grosan C (2018) Experienced gray wolf optimization through reinforcement learning and neural networks. IEEE Trans. Neural Netw. Learn. Syst. 29(3):681–694

Eswaramoorthy S, Sivakumaran N, Sekaran S (2016) Grey wolf optimization based parameter selection for support vector machines. COMPEL Int. J. Comput. Math. Electr. Electron. Eng. 35(5):1513–1523

Fasanghari M, Iranmanesh SH, Amalnick MS (2015) Predicting the success of projects using evolutionary hybrid fuzzy neural network method in early stages. J Multiple-Valued Log Soft Comput 25(2–3):291–321

Hafezi R, Shahrabi J, Hadavandi E (2015) A bat-neural network multi-agent system (BNNMAS) for stock price prediction: case study of DAX stock price. Appl Soft Comput 29:196–210

Haykin S (1994) Neural networks, a comprehensive foundation (0023527617). Macmillan, USA

Jia L, Yang B, Park DH, Tan F, Park M (2013) Software maintainability prediction model based on fuzzy neural network. J Multiple-Valued Log Soft Comput 20(1–2):39–53

Kaastra I, Boyd M (1996) Designing a neural network for forecasting financial and economic time series. Neurocomputing 10(3):215–236

Khan ZH, Alin TS, Hussain MA (2011) Price prediction of share market using artificial neural network (ANN). Int J Comput Appl 22(2):42–47

Lawrence R (1997) Using neural networks to forecast stock market prices. Univ Manitoba 333:1–21

Lertyingyod W, Benjamas N (2016) Stock price trend prediction using artificial neural network techniques case study: Thailand stock exchange. Paper presented at the 2016 computer science and engineering conference (ICSEC)

Mathworks (2012) Neural network toolbox user's guide. MATLAB, Mathworks, R2012a
Mirjalili S, Mirjalili SM, Lewis A (2014a) Let a biogeography-based optimizer train your multi-layer perceptron. Inf Sci 269:188–209
Mirjalili S, Mirjalili SM, Lewis A (2014b) Grey wolf optimizer. Adv Eng Softw 69:46–61
Navale G, Dudhwala N, Jadhav K, Gabda P, Vihangam BK (2016) Prediction of stock market using data mining and artificial intelligence. Int J Eng Sci 6(6):6539–6544
Negnevitsky M (2005) Artificial intelligence: a guide to intelligent systems. Pearson Education, USA
Patel J, Shah S, Thakkar P, Kotecha K (2015) Predicting stock and stock price index movement using trend deterministic data preparation and machine learning techniques. Expert Syst Appl 42(1):259–268
Rojas R (1996) The back-propagation algorithm. In: Neural networks. Springer, pp 149–182
Xiong T, Bao Y, Hu Z (2014) Multiple-output support vector regression with a firefly algorithm for interval-valued stock price index forecasting. Knowl-Based Syst 55:87–100
Yetis Y, Kaplan H, Jamshidi M (2014) Stock market prediction by using artificial neural network. Paper presented at the 2014 world automation congress (WAC)

Chapter 12
Future Trends in Stock Prediction Using Meta-heuristics

Abstract This chapter presents the conclusion of the book, highlighting some key findings of research. The chapter also provides current trends in stock prediction using meta-heuristics and other ensemble models. Additionally, this chapter outlines the trends that will gradually become apparent and dominate future research.

Keywords Enhancing GWO algorithm · Modified Grey Wolf Optimizer (MGWO) · Current trends in stock prediction · Meta-heuristics model · Ensemble model · Future trends in stock prediction · Meta-heuristic algorithms

12.1 Introduction

The previous chapter presented the experiments related to application of ensemble model consisting neural network and Modified Grey Wolf Optimizer (MGWO) for stock prediction. The conclusion of the book is included in this chapter that comprises some of the key findings of this research work in light of the research objectives. The contribution of this research towards knowledge is also restated in this chapter, specifically in the area of neural network and MGWO. Limitations of the research particularly the strict designing and incapability of the research are discussed as well in this chapter. The brief answers to research questions listed in Chap. 1 are also attempted in this chapter. Finally, the obstacles faced in performing the research and recommendations of future directions of this research are outlined in this chapter.

12.2 Objectives of Book Revisited

The research was aimed at enhancing the GWO algorithm and address its limitation as far as exploration and exploitation capabilities.

The objectives of the research were as below:

- To develop a modified GWO algorithm with random selection of leaders

D. Das et al., *Optimization Algorithms in Machine Learning*, Engineering Optimization: Methods and Applications, https://doi.org/10.1007/978-981-96-3849-9_12

- To adopt the modified GWO algorithm for training of neural network as ensemble model with stock market prediction analysis as case study
- To evaluate the performance of ensemble model against existing strategies in terms of the other developed optimization model in literature.

The first objective has been addressed in Chap. 3 that proposes the improvement of GWO. The GWO has been modified by strengthen the searching process via several random leaders in each iteration, re-generating the random leaders in each iteration and introducing archive to verify the solution with better probability to proceed further for training and re-generation. The improvement of exploration is achieved through the proposed approach.

The second objective has been addressed in Chap. 3 that demonstrates the design and implementation of the proposed ensemble of neural network and MGWO approach. The basis of the research is GWO algorithm that has been modified to design MGWO. The MGWO has been implemented with the wrapper's approach to select the best set of features for stock prediction. The objective has been fulfilled through the successful implementation of MGWO and neural network to form ensemble approach for stock prediction.

The third objective of the study has been achieved through the evaluation of the proposed approach, presented in Chap. 4, demonstrating the comparison of MGWO against GWO and other existing meta-heuristic approaches. The performance of ensemble approach is better than the existing strategies. For stock prediction, the performance of MGWO and ensemble approach outperforms in comparison with other existing approach as established by the statistical analysis.

Regarding the performance of MGWO and GWO, the experimental results demonstrated that MGWO outperforms GWO. Moreover, the convergence rate of MGWO is better than GWO.

Placing everything altogether, the objectives of the research has been achieved through the design, implementation and evaluation of MGWO. The research has been made significant contributions to fulfill the objectives.

12.3 Contributions of the Book

To add up the earlier discussion, the research contributes an ensemble approach in relation to stock prediction. The contribution of the research toward knowledge can be listed as:

- Modified Grey Wolf Optimizer (MGWO) modified GWO for improving the exploration of GWO via several random leaders in each iteration, re-generating the random leaders in each iteration and introducing archive to verify the solution to check the solution that provides better probability to proceed further for training and re-generation
- Ensemble of neural network and MGWO improves stock prediction with good accuracy and reduced error rate

- Ensemble approach contributes to apply it as an alternative approach for building better prediction model compared to existing approaches.

12.4 Future Directions of the Research

The research has demonstrated the modification of GWO and training neural network to form the ensemble model for stock market prediction. It has also shown the designing of algorithm applying numerous pre-defined steps that can be used for exploring stock data. The research has also illustrated the ways to reveal the useful pattern incorporated with stock data to facilitate the investor to gain through stock investment. The future research directions can be indicated as:

- Currently, the research is an attempt to modify GWO and training neural network to form ensemble model for stock market prediction. To improve and strengthen the result of the research, it is possible to carry out better ensemble or hybrid approach.
- The study investigated to implement ensemble approach applying MLP neural network and MGWO. Further investigations can be made to combine other methods and various parameters can be fine-tuned for better improvement.
- The improvement of computational cost for MGWO algorithm may be investigated by implementing the algorithm for parallel approach. The complexity of the MGWO algorithm may be addressed by fine tuning the algorithm.
- This study uses daily stock data for six years of stock data. However, attempts can be made to consider the stock data with different time period. Additionally, different ratio of data may be considered for training, validation and testing.
- Positive and negative news can be incorporated with stock data to measure their effect on stock movement.
- Moreover, the research attempted to measure the effect of five different factors on stock price of DSE, which may be extended for more factors. In addition, attempt can be made further to collect the factors data for other stock markets.
- While predicting the stock market, clustering is an important and complicated task that can be solved building ensemble model consisting metaheuristic algorithms like GWO-KMeans, PSO-KMeans, BAT-KMeans, Firefly-KMeans, and so on.
- Meta-heuristic algorithms have the capability to provide solution for multidisciplinary problems. Feature selection is another important aspect of prediction to obtain better accuracy and meta-heuristic algorithms based ensemble model can be built to address this.
- Integration of Meta-heuristic and machine learning can be performed to build an ensemble model and this model can be applied for solving real world optimization problem. This could be another aspect that needs further investigation.
- The optimization of Deep learning architecture can be investigated by applying meta-heuristic algorithms like GWO, PSO, BBO and firefly. Besides, ensemble model can be built through meta-heuristic algorithms for the optimization of artificial neural network (ANN) internal weights as well as hyperparameters like

batch size, activation function, number of layers, learning rate, number of epochs, etc.

- Finally, the performance of MGWO algorithm may be fine-tuned through hybrid it with other meta-heuristic algorithm so that the algorithm may gain better search ability.

12.5 Current Trends in Stock Prediction Using Meta-heuristics and Other Ensemble Models

Over the years, Data Science and Artificial Intelligence evolving so rapidly that shapes the meta-heuristic in better form by proposing new model continuously in one side and hence the stock prediction is also getting better day by day other side. Here, we will try to explore some of the recent works attempted by the researchers leveraging the ensemble model for stock market prediction. Some of the significant research attempted in this area are, but not limited to:

- Metaheuristic enabled ensembled intelligent model can be an excellent approach for stock prediction. An ensemble of Metaheuristic enabled intelligent model for stock market prediction called FU-WOA–NN has been proposed by Tripathi et al. (2023) consisting of a Neural Network (NN) and Fly Updated Whale Optimization Algorithm (FU-WOA). Their model outperformed the other meta-heuristic based ensemble and hybrid algorithm for predicting the stock market of India and other Asian countries.
- Sentiment analysis of stock market is a promising dimension that could be explored deploying natural language processing to improve stock prediction. Colasanto et al. (2022) investigated sentiment analysis model and applied sentiment polarity score to improve the stock prediction. This research accomplished encouraging results and hence, further research can be undertaken towards this direction.
- Integration of Transform model for feature engineering and machine learning model for predicting stock price is another decent model for stock market. An attempt of Hilbert–Huang Transform and XGBoost Classifier for the prediction of S&P500 stocks in US Stock market has been experimented by Dezhkam and Manzuri (2023), where Hilbert–Huang Transform (HHT) is leveraged for the feature engineering and the extreme gradient boost (XGBoost) is employed for the Close price trend of stock market. The outcome of this research is pretty promising and it can perform well even with the poor-performing markets.
- Integration of multiple models can bring favorable results for most of the prediction including stock market. In light of this approach, LASSO–TLBO–SVR has been experimented by Mishra et al. (2021) for portfolio construction and it has been experimented for the return of the Indian market portfolio (NSE and BSE). The outcome is pretty pleasing for this study and the model can be really promising for the market portfolio.

- Another good option for the prediction of stock market can be Integration of various deep learning techniques. Song et al. (2023) proposed the similar techniques consist of the Recurrent Neural Network Based Hybrid Models, CNN-LSTM, GRU-CNN, and Ensemble Models for forecasting Stock Market Indices. The model predicted with lower error rate for one-time-step and multi-time-step predictions of the closing price for three different stock market indices in different financial markets.
- Improvement of stock market prediction can be made by adopting an ensemble approach leveraging meta-heuristic algorithms and time series model. A similar approach has been explored by Shahvaroughi Farahani et al. (2023) that exhibited promising result combining genetic algorithm (GA) and a hybrid of grey wolf optimization and particle swarm optimization binary algorithm (GWO-PSO) for feature selection and harmony search algorithm (HS), particle swarm optimization algorithm (PSO), modified particle swarm optimization algorithm (MPSO), modified particle swarm optimization algorithm with time-varying acceleration coefficients (MPSO-TVAC), moth flame optimization (MFO), wolf optimization algorithm (WOA) and chimp optimization algorithm (ChOA) for training neural network.
- Combination of Supervised Machine Learning Algorithm like Support Vector Machine (SVM) and Metaheuristic Algorithms can be an effective approach for stock predication. In this regard, Mahmoodi et al. (2023) attempted to combine SVM and particle swarm optimization (PSO) to produce a promising stock market timing model.
- Optimizing the parameter of Least Squares Support Vector Machine (LS-SVM) by integrating it with the Meta Heuristic algorithm is another wonderful approach to predict the future stock price. Chatterjee et al. (2023) proposed a similar approach integrating LS-SVM and Altruistic Dragonfly Algorithm (ADA) for the improvement of stock market prediction performance.
- Combination of Deep Structured Learning methods can be a great option for portfolio forecasting of stock market. Sharma and Shekhawat (2022) inspired to adopt similar approach by integrating Deep Belief Network (DBN) with Recurrent Neural Network (RNN) that brought the realistic result in terms of return and risk of the stock market.
- Integration of Support Vector Machine (SVM) with metaheuristic algorithm can also be considered as one of the better options for stock prediction. A similar study conducted by Mahmoodi et al. (2023) provided good stock price prediction capability, where they have combined SVM and PSO. The obtained result is also better than the integration of other algorithms like SVM-CS and similar integration.
- Optimization of LSTM through Metaheuristic Algorithm like Neural Network Algorithm for the prediction of financial time series can be another worthy alternative that has been managed to perform a notable improvement of 40, 65, 4, and 85% in the MAPE, Theil U, R, and RMSE metrics for stock market experimented by Dastgerdi and Mercorelli (2023). Eventually, this work produced more accurate and precise prediction results.

- Improvement of stock intra-day prediction through the optimization of meta-heuristic algorithm such as flower pollination algorithm and particle swarm optimization combined with RNN-LSTM can be an excellent approach. Kumar and Haider (2021) attempted similar approach that could enhance the stock prediction to attain maximum forecasting accuracy approximately increment of 4–6% through the implementation of the metaheuristic approach and integration.

12.6 Future Trends in Stock Prediction Research

Increasing the stock investment returns through effective stock price prediction trends will remain an important problem for financial markets. Consequently, enhancement of algorithmic performance through optimization and similar approaches for stock predictions will remain center of research even in future (Islam et al. 2021). Thus, this section outlines the trends that will gradually become apparent even in future and dominate future research in order to further enhance the state-of-the-art pertinent to this knowledge area.

- For future research, the stock prediction system can be improved through adopting alternative metaheuristic algorithms. Moreover, this type of prediction system can be integrated into a dynamic trading system or an automated stock market system. It can be aimed to implement the successful metaheuristic models in the dynamic trading system, so that the future values of the stocks can be predicted with low error and the trading system can provide successful buy/sell/keep suggestions.
- Developing new metaheuristic algorithms can be another good option to improve the accuracy of stock prediction and reduce error rate. As the focus is different for different algorithms like the focus for some algorithms can be on using populations and imitating social behavior; some focus on adjusting the parameter settings dynamically; and others focus on applying metaheuristic algorithms to parallel computing systems. Hence, the performance of each algorithm may vary. Another efficient metaheuristic algorithm known as Spider monkey optimization algorithm that is inspired by social behavior of spider monkeys can be experimented for stock prediction as the algorithm has demonstrated its potential to various problem solving like optimization, numerical problems, image recognition processing, local search problem, traffic delay problem, energy-efficient clustering, and communication optimization (Tsai et al. 2019).
- Traditional optimization algorithms have the limitation of getting stuck in a local minimum and hence those may not be able to find the best solution. However, metaheuristic algorithms can escape from local minima and capable to determine the best solution across the entire search space. Moreover, integration of meta-heuristic algorithms is beneficial to handle complex and noisy data. Another novel algorithm known as Artificial rabbits optimization (ARO) proposed by Wang et al. (2022) can be tested for stock prediction with the integration of other AI based algorithms.

- Technical indicators of stock prediction can be extended in various ways. Stock data can be analyzed at hourly, minute or second intervals and then different selection algorithms can be leveraged through multi-criteria decision making. Moreover, various time series algorithms and integration of different algorithms can be applied. As stock markets are highly uncertain, prediction is extremely difficult. Thus, latest and powerful metaheuristic algorithms such as Lion-Inspired, Red Deer, Social Engineering Optimization for model parameters can be experimented to reveal such stock prediction complexity and uncertainty.

12.7 Suggestions for Future Work

Selection of the best model for stock prediction is a debatable issue and hence continuous research is required to resolve this issue. Moreover, researchers need continuous concentration on developing and evaluating new models. Additional efforts and more researches are required for data extraction and synthesis, especially from multiple systematic reviews. Another important aspect is reviewing the practice and policy, where the researchers can investigate how decision making can be made through better model and what factors affect this. As integration of models and ensemble approaches have demonstrated significant potential, it is better to investigate more on this type of integration techniques.

Overall, metaheuristic artificial neural network based models like SVM, LSTM, NN, ANN, CNN, RNN, and their ensemble or hybrid versions are some of the most frequently applied models for stock prediction. The most leveraged information sources are the time series of historical closing stock prices, the most used performance metrics of the predictive model is accuracy and in future, it would be better to employ and integrate as more information and technologies as possible for the stock price prediction. Consequently, these new findings would certainly facilitate improving the accuracy of stock predictions.

Besides, it is recommended for the future meta-heuristics based stock prediction research to utilize complex models consisting multiple constraints motivated by the real world. Furthermore, it is always better to examine the performance of the algorithms in terms of real world than the close solutions determined by the algorithms. As it is really challenging to determine the concrete meta-heuristic algorithm for stock prediction, the optimization of various meta-heuristic algorithms will remain an attractive research area for researchers. Thus, similar work and their extensions should be continuously investigated to unleash the significant findings. Apart from that, various researches suggest huge number of meta-heuristic algorithms, their hybrid and enhanced forms. It will be better to experiment those algorithms for the stock prediction problem.

Finally, future studies should investigate various novel meta-heuristic algorithms like Aquila Optimization Algorithm (AO), Bald Eagle Search (BES) optimization algorithms, Squid Game Optimizer (SGO), Geometric Mean Optimizer (GMO),

Victoria Amazonica Optimization (VAO) algorithm, Pelican Optimization Algorithm, White Shark Optimizer (WSO), Wild Horse Optimizer (WHO), Elephant Herding Optimization (EHO), Galactic Swarm Optimization (GSO), etc. for solving the non-linearity of stock prediction.

References

Chatterjee B, Acharya S, Bhattacharyya T, Mirjalili S, Sarkar R (2023) Stock market prediction using altruistic dragonfly algorithm. PLoS ONE 18(4):e0282002

Colasanto F, Grilli L, Santoro D, Villani G (2022) BERT's sentiment score for portfolio optimization: a fine-tuned view in Black and Litterman model. Neural Comput Appl 34(20):17507–17521

Dastgerdi AK, Mercorelli P (2023) Employing a novel metaheuristic algorithm to optimize an LSTM model: a case study of stock market prediction. In: International conference on advances in computing and data sciences, Apr 2023. Springer Nature Switzerland, Cham, pp 478–489

Dezhkam A, Manzuri MT (2023) Forecasting stock market for an efficient portfolio by combining XGBoost and Hilbert–Huang transform. Eng Appl Artif Intell 118:105626

Islam S, Sikder MS, Hossain MF, Chakraborty P (2021) Predicting the daily closing price of selected shares on the Dhaka stock exchange using machine learning techniques. SN Bus Econ 1(4):58

Kumar K, Haider MTU (2021) Enhanced prediction of intra-day stock market using metaheuristic optimization on RNN–LSTM network. New Gener Comput 39(1):231–272

Mahmoodi A, Hashemi L, Jasemi M, Mehraban S, Laliberté J, Millar RC (2023) A developed stock price forecasting model using support vector machine combined with metaheuristic algorithms. Opsearch 60(1):59–86

Mishra S, Padhy S, Mishra SN, Misra SN (2021) A novel LASSO–TLBO–SVR hybrid model for an efficient portfolio construction. N Am J Econ Financ 55:101350

Shahvaroughi Farahani M, Farrokhi-Asl H, Rahimian S (2023) Hybrid metaheuristic artificial neural networks for stock price prediction considering efficient market hypothesis. Int J Res Ind Eng 12(3):234–272

Sharma M, Shekhawat HS (2022) Portfolio optimization and return prediction by integrating modified deep belief network and recurrent neural network. Knowl-Based Syst 250:109024

Song M, Li Y, Pedrycz W (2023) Time series prediction with granular neural networks. Neurocomputing 546:126328

Tripathi DK, Chadha S, Tripathi A (2023) Metaheuristic enabled intelligent model for stock market prediction via integrating volatility spillover: India and its Asian and European counterparts. Data Knowl Eng 144:102127

Tsai CW, Ding YC, Liu SJ, Chiang MC, Yang CS (2019) A high-performance clustering algorithm based on searched experiences. Comput Hum Behav 100:231–241

Wang L, Cao Q, Zhang Z, Mirjalili S, Zhao W (2022) Artificial rabbits optimization: a new bio-inspired meta-heuristic algorithm for solving engineering optimization problems. Eng Appl Artif Intell 114:105082

Appendix
Stock Datasets

Stock Dataset

Unprocessed Stock Data from Bursa Malaysia (Partial)

Stock number	Stock name	Date	Open price	High price	Low price	Closing price	Volume
6947	Digi Com	20050103	0.5525	0.5525	0.5347	0.5436	728,000
		20050104	0.5392	0.5436	0.5392	0.5392	1,981,000
		20050105	0.5392	0.5436	0.5392	0.5436	2,178,000
		20050106	0.5436	0.5436	0.5392	0.5436	1,880,000
		20050107	0.5436	0.5481	0.5392	0.5392	5,528,000
		20050110	0.5392	0.5525	0.5392	0.5392	7,693,000
		20050111	0.5392	0.5481	0.5347	0.5436	4,614,000
		20050112	0.5436	0.5481	0.5436	0.5481	3,167,000
		20050113	0.5481	0.5525	0.5436	0.5436	4,632,000
		20050114	0.5525	0.5525	0.5436	0.5436	3,392,000
		20050117	0.5481	0.557	0.5436	0.5481	6,807,000
		20050118	0.5481	0.5481	0.5436	0.5436	7,070,000
		20050119	0.5525	0.5525	0.5392	0.5392	1,384,000
		20050120	0.5436	0.5436	0.5392	0.5392	4,064,000
		20050124	0.5392	0.5392	0.5347	0.5347	2,525,000
		20050125	0.5347	0.5392	0.5258	0.5302	4,994,000
		20050126	0.5302	0.5302	0.5124	0.5258	7,393,000
		20050127	0.5213	0.5258	0.5213	0.5213	1,591,000
		20050128	0.5213	0.5213	0.5169	0.5213	1,458,000
		20050131	0.5213	0.5347	0.508	0.508	7,347,000
		20050202	0.5124	0.5169	0.508	0.5124	3,057,000

(continued)

D. Das et al., *Optimization Algorithms in Machine Learning*, Engineering Optimization: Methods and Applications, https://doi.org/10.1007/978-981-96-3849-9

(continued)

Stock number	Stock name	Date	Open price	High price	Low price	Closing price	Volume
		20050203	0.5124	0.5347	0.5124	0.5258	15,602,000
		20050204	0.5213	0.5302	0.5213	0.5258	4,731,000
		20050207	0.5302	0.5347	0.5302	0.5302	8,840,000
		20050208	0.5347	0.5392	0.5302	0.5347	8,257,000
		20050214	0.5392	0.5481	0.5347	0.5436	27,897,000
		20050215	0.5481	0.5525	0.5436	0.5436	17,165,000
		20050216	0.5436	0.5481	0.5436	0.5481	12,852,000
		20050217	0.5436	0.5481	0.5347	0.5347	4,589,000
		20050218	0.5347	0.5347	0.5347	0.5347	3,629,000
		20050221	0.5302	0.5347	0.5302	0.5302	4,545,000
		20050222	0.5302	0.5347	0.5258	0.5347	5,380,000
		20050223	0.5258	0.5258	0.5169	0.5213	2,268,000
		20050224	0.5258	0.5258	0.5169	0.5213	1,489,000

Unprocessed Stock Data from DSE, Bangladesh (Partial)

Date	Company code	Low	High	Average	Close	Trd. vol.	Turn over	Company name
2011-01-02	18455	366.00	380.00	373.02	372.82	3,467,205.00	19,404,000.00	ACI
2011-01-03	18455	360.00	372.80	369.89	370.17	2,456,430.00	19,404,000.00	
2011-01-04	18455	365.00	373.00	367.10	366.62	3,941,125.00	19,404,000.00	
2011-01-05	18455	350.00	371.00	366.09	367.10	3,046,935.00	19,404,000.00	
2011-01-06	18455	345.00	365.10	361.24	361.82	3,618,185.00	19,404,000.00	
2011-01-09	18455	333.20	361.00	343.41	344.03	3,478,150.00	19,404,000.00	
2011-01-10	18455	303.10	335.00	316.37	316.48	1,155,140.00	19,404,000.00	
2011-01-11	18455	325.00	364.80	359.10	359.29	10,228,915.00	19,404,000.00	
2011-01-12	18455	336.60	357.10	351.20	351.76	4,783,995.00	19,404,000.00	
2011-01-13	18455	345.00	353.80	348.07	348.14	3,376,940.00	19,404,000.00	
2011-01-16	18455	331.00	365.00	342.16	345.23	6,645,730.00	19,404,000.00	
2011-01-17	18455	340.00	350.00	345.37	345.40	6,139,820.00	19,404,000.00	
2011-01-18	18455	305.00	344.00	338.08	338.92	2,018,931.10	19,404,000.00	
2011-01-19	18455	320.00	336.10	327.80	327.65	2,080,605.00	19,404,000.00	
2011-01-20	18455	310.00	323.00	316.67	319.67	383,600.00	19,404,000.00	
2011-01-25	18455	320.10	347.50	343.49	343.38	7,708,890.00	19,404,000.00	
2011-01-26	18455	340.00	355.00	346.70	347.07	7,845,280.00	19,404,000.00	
2011-01-27	18455	335.00	350.00	343.49	343.87	5,147,720.00	19,404,000.00	
2011-01-30	18455	325.00	343.90	340.64	340.59	1,631,440.00	19,404,000.00	
2011-01-31	18455	315.00	340.00	337.95	338.32	4,976,750.00	19,404,000.00	
2011-02-01	18455	316.00	340.00	334.77	334.50	2,953,655.00	19,404,000.00	
2011-02-02	18455	320.00	333.40	329.87	329.83	2,558,795.00	19,404,000.00	

(continued)

(continued)

Date	Company code	Low	High	Average	Close	Trd. vol.	Turn over	Company name
2011-02-03	18455	311.00	327.60	323.61	323.55	2,482,265.00	19,404,000.00	
2011-02-06	18455	302.00	320.00	309.91	309.95	9,546,580.00	19,404,000.00	
2011-02-07	18455	285.00	310.00	303.59	303.87	7,226,035.00	19,404,000.00	
2011-02-08	18455	298.00	315.00	304.81	303.54	7,876,930.00	19,404,000.00	
2011-02-09	18455	308.00	330.00	314.72	314.45	3,179,125.00	19,404,000.00	
2011-02-10	18455	295.00	307.90	301.11	300.98	2,531,230.00	19,404,000.00	
2011-02-13	18455	271.30	290.20	275.79	275.59	5,635,885.00	19,404,000.00	
2011-02-14	18455	248.00	270.00	248.79	248.67	3,817,105.00	19,404,000.00	
2011-02-15	18455	227.00	260.00	243.96	242.72	5,162,555.00	19,404,000.00	
2011-02-20	18455	259.00	265.00	264.48	264.67	1,805,040.00	19,404,000.00	

Unprocessed Stock Data from NYSE, USA (Partial)

Company name	Date	Open	High	Low	Close	Volume
IBM, NYSE	3-Jan-11	147.21	148.2	147.14	147.48	4,603,800
	4-Jan-11	147.56	148.22	146.64	147.64	5,060,100
	5-Jan-11	147.34	147.48	146.73	147.05	4,657,400
	6-Jan-11	147.13	148.79	146.82	148.66	5,029,200
	7-Jan-11	148.79	148.86	146.94	147.93	4,135,700
	10-Jan-11	147.58	148.06	147.23	147.64	3,633,400
	11-Jan-11	148.2	148.35	146.75	147.28	4,163,600
	12-Jan-11	147.99	149.29	147.67	149.1	4,091,500
	13-Jan-11	149.24	149.29	148.25	148.82	3,445,800
	14-Jan-11	148.89	150	148.47	150	4,544,200
	18-Jan-11	149.82	151.46	149.38	150.65	9,176,900
	19-Jan-11	153.26	156.13	152.83	155.69	12,141,000
	20-Jan-11	154.53	155.96	154.45	155.8	7,439,900
	21-Jan-11	156.4	156.78	154.96	155.5	7,009,000
	24-Jan-11	155.42	159.79	155.33	159.63	7,285,100
	25-Jan-11	159.21	164.35	159	161.44	8,260,800
	26-Jan-11	161.67	161.9	160.42	161.04	5,353,100
	27-Jan-11	161.43	162.18	160.86	161.07	4,878,300
	28-Jan-11	161.05	161.92	158.67	159.21	6,725,600
	31-Jan-11	159.18	162	158.68	162	7,197,200
	1-Feb-11	162.11	163.94	162	163.56	5,831,300
	2-Feb-11	163.4	163.6	162.61	163.3	3,904,000
	3-Feb-11	163.16	164.2	162.81	163.53	4,683,400
	4-Feb-11	163.48	164.14	163.22	164	3,755,200
	7-Feb-11	164.08	164.99	164.02	164.82	4,928,100
	8-Feb-11	164.82	166.25	164.32	166.05	5,612,600
	9-Feb-11	165.62	165.97	164.1	164.65	4,633,600
	10-Feb-11	163.9	165	163.18	164.09	5,737,800
	11-Feb-11	163.98	165.01	163.31	163.85	5,185,200
	14-Feb-11	164.18	164.38	162.85	163.22	4,129,800
	15-Feb-11	162.89	163.57	162.52	162.84	3,768,700
	16-Feb-11	163.33	163.6	162.75	163.4	3,216,000
	17-Feb-11	163.3	164.67	162.85	164.24	3,230,500
	18-Feb-11	164.46	164.84	164.1	164.84	4,245,000
	22-Feb-11	163.57	164.26	161.78	161.95	5,209,300
	23-Feb-11	161.81	162.68	160.14	160.18	5,998,100

Unprocessed Stock Data from NASDAQ, USA (Partial)

Company name	Date	Open	High	Low	Close	Volume
Microsoft, NASDAQ	3-Jan-11	28.05	28.18	27.92	27.98	53,443,800
	4-Jan-11	27.94	28.17	27.85	28.09	54,405,600
	5-Jan-11	27.9	28.01	27.77	28	58,998,700
	6-Jan-11	28.04	28.85	27.86	28.82	88,026,300
	7-Jan-11	28.64	28.74	28.25	28.6	73,762,000
	10-Jan-11	28.26	28.4	28.04	28.22	57,573,600
	11-Jan-11	28.2	28.25	28.05	28.11	50,298,900
	12-Jan-11	28.12	28.59	28.07	28.55	52,631,100
	13-Jan-11	28.33	28.39	28.01	28.19	67,077,600
	14-Jan-11	28.08	28.38	27.91	28.3	62,688,400
	18-Jan-11	28.16	28.74	28.14	28.66	53,322,700
	19-Jan-11	28.46	28.68	28.27	28.47	50,005,900
	20-Jan-11	28.5	28.55	28.13	28.35	58,613,600
	21-Jan-11	28.4	28.43	28.02	28.02	58,080,300
	24-Jan-11	28.02	28.56	27.99	28.38	52,047,800
	25-Jan-11	28.14	28.45	28.12	28.45	42,436,600
	26-Jan-11	28.51	28.99	28.5	28.78	74,628,800
	27-Jan-11	28.75	29.46	28.49	28.87	146,938,600
	28-Jan-11	28.9	28.93	27.45	27.75	141,249,400
	31-Jan-11	27.77	27.9	27.42	27.73	65,029,000
	1-Feb-11	27.8	28.06	27.61	27.99	62,810,700
	2-Feb-11	27.93	28.11	27.88	27.94	45,824,000
	3-Feb-11	27.97	27.97	27.54	27.65	60,340,100
	4-Feb-11	27.7	27.84	27.51	27.77	40,412,200
	7-Feb-11	27.8	28.34	27.79	28.2	68,980,900
	8-Feb-11	28.1	28.34	28.05	28.28	34,904,200
	9-Feb-11	28.19	28.26	27.91	27.97	52,905,100
	10-Feb-11	27.93	27.94	27.29	27.5	76,672,400
	11-Feb-11	27.76	27.81	27.07	27.25	83,939,700
	14-Feb-11	27.21	27.27	26.95	27.23	56,766,200
	15-Feb-11	27.04	27.33	26.95	26.96	44,116,500
	16-Feb-11	27.05	27.07	26.6	27.02	70,817,900
	17-Feb-11	26.97	27.37	26.91	27.21	57,207,300
	18-Feb-11	27.13	27.21	26.99	27.06	68,667,800
	22-Feb-11	26.78	27.1	26.52	26.59	60,889,000
	23-Feb-11	26.53	26.86	26.43	26.59	60,234,100

(continued)

(continued)

Company name	Date	Open	High	Low	Close	Volume

Formatted Stock Data for Pharmaceutical Sector (Partial) Used for k-Means Clustering

Company name	Date	Low	High	Average	Close	Trd. vol.	Turn over
ACI	2011-01-02	366	380	373.02	372.9	979,300	3,467,205
ACI	2011-01-03	360	372.8	369.89	370.7	706,636	2,456,430
ACI	2011-01-04	365	373	367.1	366.6	9,310,750	3,941,125
ACI	2011-01-05	350	371	366.09	369	878,300	3,046,935
ACI	2011-01-06	345	365.1	361.24	357.7	10,110,000	3,618,185
ACI	2011-01-09	333.2	361	343.41	340.1	10,810,110	3,478,150
ACI	2011-01-10	303.1	335	316.37	311.3	493,650	1,155,140
BXPHARMA	2011-01-02	110	145	139.66	141.7,3020	1,025,331	143,251,680.2
BXPHARMA	2011-01-03	114	148	140.74	139.8,1718	569,133	80,339,041.5
BXPHARMA	2011-01-04	112	153	138.1	137.3,1360	412,050	56,988,008.2
BXPHARMA	2011-01-05	110	140	135.61	135.6,1149	383,854	52,136,569
BXPHARMA	2011-01-06	109.1	142	135.82	133.3,1767	667,923	90,841,559.9
BXPHARMA	2011-01-09	115	136	131.35	126.7,1948	792,768	104,646,626.6
BXPHARMA	2011-01-10	110	130	122.46	119.6,457	186,421	22,919,545
GLAXOSMITH	2011-01-02	1121.1	1152	1142.24	1141.6	382,700	3,084,655
GLAXOSMITH	2011-01-03	1140	1168	1148.42	1143.7	533,600	4,131,570
GLAXOSMITH	2011-01-04	1110	1147	1123.2	1117.7	252,200	2,464,695
GLAXOSMITH	2011-01-05	1105	1129	1118.3	1111.1	512,800	3,132,765
GLAXOSMITH	2011-01-06	1080	1120	1105.91	1104.8	352,350	2,599,745
GLAXOSMITH	2011-01-09	970	1101	1018.07	998.4	493,550	3,591,410
GLAXOSMITH	2011-01-10	932	990	954.33	953.5	6400	381,400
RECKITBEN	2011-01-02	1207.1	1290	1238.42	1238.9	10,550	681,410

(continued)

(continued)

Company name	Date	Low	High	Average	Close	Trd. vol.	Turn over
RECKITBEN	2011-01-03	1214	1244	1222.65	1222.4	14,750	916,855
RECKITBEN	2011-01-04	1200	1215	1209.01	1209	8400	483,605
RECKITBEN	2011-01-05	1195	1257	1235.99	1235.9	201,000	1,235,990
RECKITBEN	2011-01-06	1201	1230	1209.26	1208.6	12,650	785,605
RECKITBEN	2011-01-09	1058.1	1150	1098.2	1098.2	10,500	549,100
SQURPHARMA	2011-01-02	3521	3565	3553.74	3559.75,2349	33,997	120,842,594
SQURPHARMA	2011-01-03	3525.25	3580	3553.84	3534.5,2609	26,008	92,456,295
SQURPHARMA	2011-01-04	3500.25	3549	3515.99	3507.25,2215	24,329	85,563,216.5
SQURPHARMA	2011-01-05	3464	3534.75	3482.11	3478.5,2900	20,678	72,001,794.75
SQURPHARMA	2011-01-06	3403	3505	3440.84	3415.75,2652	20,058	69,006,406.5
SQURPHARMA	2011-01-09	3145	3450	3281.79	3176.25,4904	73,526	238,928,996.5
SQURPHARMA	2011-01-10	3010	3269	3063.89	3030,1309	28,409	86,424,439.75

Factors Data (Partial)

Gold price	Dollar price	Bank interest rate	FDI	Inflation
2905.00	69.25	8.50	861,736,237.16	9.00
2905.00	69.25	8.50	861,736,237.16	9.00
2905.00	69.24	8.50	861,736,237.16	9.00
2905.00	69.24	8.50	861,736,237.16	9.00
2905.00	69.24	8.50	861,736,237.16	9.00
2905.00	69.27	8.50	861,736,237.16	9.00
2905.00	69.26	8.50	861,736,237.16	9.00
2905.00	69.25	8.50	861,736,237.16	9.00
2905.00	69.26	8.50	861,736,237.16	9.00
2905.00	69.25	8.50	861,736,237.16	9.00
2905.00	69.25	8.50	861,736,237.16	9.00
2905.00	69.25	8.50	861,736,237.16	9.00
2905.00	69.25	8.50	861,736,237.16	9.00
2905.00	69.25	8.50	861,736,237.16	9.00
2905.00	69.25	8.50	861,736,237.16	9.00
2905.00	69.25	8.50	861,736,237.16	9.00
2905.00	69.25	8.50	861,736,237.16	9.00
2905.00	69.25	8.50	861,736,237.16	9.00
2905.00	69.25	8.50	861,736,237.16	9.00
2995.00	69.25	8.50	861,736,237.16	9.00
2995.00	69.26	8.50	861,736,237.16	9.00
2995.00	69.31	8.50	861,736,237.16	9.00
2995.00	69.30	8.50	861,736,237.16	9.00
2995.00	69.30	8.50	861,736,237.16	9.00
2995.00	69.31	8.50	861,736,237.16	9.00
2995.00	69.31	8.50	861,736,237.16	9.00
2995.00	69.29	8.50	861,736,237.16	9.00
2995.00	69.29	8.50	861,736,237.16	9.00
2995.00	69.29	8.50	861,736,237.16	9.00
2995.00	69.30	8.50	861,736,237.16	9.00
3095.00	69.27	8.50	861,736,237.16	9.00
3095.00	69.27	8.50	861,736,237.16	9.00
3095.00	69.27	8.50	861,736,237.16	9.00
3095.00	69.27	8.50	861,736,237.16	9.00
3095.00	69.29	8.50	861,736,237.16	9.00
3095.00	69.31	8.50	861,736,237.16	9.00
3095.00	69.33	8.50	861,736,237.16	9.00

(continued)

(continued)

Gold price	Dollar price	Bank interest rate	FDI	Inflation
3095.00	69.38	8.50	861,736,237.16	9.00
3095.00	69.39	8.50	861,736,237.16	9.00

Formatted Stock Data for Microsoft, NASDAQ (Partial) Used for Classification

Open	High	Low	Close	Volume	Decision
37.35	37.4	37.1	37.16	30,632,200	1
37.2	37.22	36.6	36.91	31,134,800	2
36.85	36.89	36.11	36.13	43,603,700	2
36.33	36.49	36.21	36.41	35,802,800	1
36	36.14	35.58	35.76	59,971,700	2
35.88	35.91	35.4	35.53	36,516,300	2
35.9	36.15	35.75	36.04	40,548,800	3
35.99	36.02	34.83	34.98	45,901,900	1
34.73	35.88	34.63	35.78	41,623,300	1
35.9	36.79	35.85	36.76	44,812,600	3
36.69	37	36.31	36.89	38,018,700	3
36.83	36.83	36.15	36.38	46,267,500	1
36.82	36.82	36.06	36.17	31,567,300	2
36.26	36.32	35.75	35.93	21,904,300	2
36.09	36.13	35.52	36.06	43,954,000	1
37.45	37.55	36.53	36.81	76,395,500	3
36.87	36.89	35.98	36.03	44,420,800	2
36.12	36.39	35.75	36.27	36,205,500	1
35.98	36.88	35.9	36.66	52,745,900	1
36.79	36.88	36.23	36.86	35,036,300	1
36.95	37.89	36.56	37.84	93,162,300	3
37.74	37.99	36.43	36.48	64,063,100	1
36.97	37.19	36.25	36.35	54,697,900	2
36.29	36.47	35.8	35.82	55,814,400	2
35.8	36.25	35.69	36.18	35,351,800	1
36.32	36.59	36.01	36.56	33,260,500	3
36.63	36.8	36.29	36.8	26,767,000	3
36.88	37.26	36.86	37.17	32,141,400	3
37.35	37.6	37.3	37.47	27,051,800	3
37.33	37.86	37.33	37.61	37,635,500	1
37.39	37.78	37.33	37.62	31,407,500	1

(continued)

(continued)

Open	High	Low	Close	Volume	Decision
37.63	37.78	37.41	37.42	32,834,000	1
37.22	37.75	37.21	37.51	29,750,400	1
37.57	37.87	37.4	37.75	27,526,100	3
37.94	38.35	37.86	37.98	38,021,300	3
37.69	37.98	37.54	37.69	32,085,100	2
37.61	37.85	37.35	37.54	30,736,500	2

Normalized Stock Data with Factors for DSE (Partial) Used for Classification and Prediction

Low	High	Average	Close	Trd. vol.	Turn over	Gold dollar	Bank interest	FDI	Inflation	Decision
0.64	0.65	0.64	0.64	0.03	0.31	0.11	0	0.5	0.6	1
0.63	0.63	0.63	0.63	0.02	0.31	0.11	0	0.5	0.6	2
0.64	0.63	0.62	0.62	0.03	0.31	0.11	0.38	0.5	0.6	1
0.6	0.63	0.62	0.62	0.03	0.31	0.11	0.38	0.5	0.6	1
0.59	0.61	0.61	0.61	0.03	0.31	0.11	0.38	0.5	0.6	1
0.56	0.6	0.56	0.56	0.03	0.31	0.12	0.38	0.5	0.6	1
0.48	0.53	0.49	0.49	0.01	0.31	0.12	0.38	0.5	0.6	2
0.53	0.61	0.6	0.6	0.09	0.31	0.12	0.38	0.5	0.6	3
0.57	0.59	0.58	0.58	0.04	0.31	0.12	0.38	0.5	0.6	1
0.59	0.58	0.57	0.57	0.03	0.31	0.12	0.38	0.5	0.6	1
0.55	0.61	0.56	0.56	0.06	0.31	0.12	0.38	0.5	0.6	1
0.57	0.57	0.57	0.56	0.06	0.31	0.12	0.38	0.5	0.6	1
0.48	0.55	0.55	0.55	0.02	0.31	0.12	0.38	0.5	0.6	2
0.52	0.53	0.52	0.52	0.02	0.31	0.12	0.38	0.5	0.6	1
0.49	0.5	0.49	0.5	0	0.31	0.12	0.38	0.5	0.6	2
0.52	0.56	0.56	0.56	0.07	0.31	0.13	0.88	0.5	0.6	1
0.57	0.58	0.57	0.57	0.07	0.31	0.13	0.88	0.5	0.6	3
0.56	0.57	0.56	0.56	0.05	0.31	0.13	0.88	0.5	0.6	2
0.53	0.55	0.55	0.55	0.01	0.31	0.13	0.88	0.5	0.6	2
0.51	0.54	0.55	0.55	0.05	0.27	0.13	0.88	0.5	0.6	1
0.51	0.54	0.54	0.54	0.03	0.27	0.13	0.88	0.5	0.6	1
0.52	0.53	0.52	0.52	0.02	0.27	0.13	0.88	0.5	0.6	1
0.5	0.51	0.51	0.51	0.02	0.27	0.13	0.88	0.5	0.6	2
0.47	0.49	0.47	0.47	0.1	0.27	0.13	0.88	0.5	0.6	1
0.43	0.46	0.45	0.45	0.08	0.27	0.13	0.88	0.5	0.6	2

(continued)

(continued)

Low	High	Average	Close	Trd. vol.	Turn over	Gold dollar	Bank interest	FDI	Inflation	Decision
0.46	0.48	0.46	0.45	0.08	0.27	0.13	0.88	0.5	0.6	1
0.49	0.52	0.48	0.48	0.03	0.27	0.13	0.88	0.5	0.6	1
0.46	0.46	0.45	0.44	0.03	0.27	0.13	0.88	0.5	0.6	2
0.39	0.41	0.38	0.38	0.07	0.27	0.13	0.88	0.5	0.6	1
0.33	0.36	0.31	0.3	0.05	0.27	0.13	0.88	0.5	0.6	2
0.28	0.33	0.29	0.29	0.07	0.27	0.13	0.88	0.5	0.6	1
0.36	0.34	0.35	0.35	0.02	0.27	0.13	0.88	0.5	0.6	1
0.31	0.41	0.4	0.4	0.14	0.27	0.13	0.88	0.5	0.6	1
0.39	0.38	0.38	0.37	0.03	0.31	0.13	0.88	0.5	0.6	1
0.34	0.39	0.37	0.37	0.04	0.31	0.13	0.88	0.5	0.6	1
0.32	0.34	0.31	0.31	0.06	0.31	0.13	0.88	0.5	0.6	1
0.28	0.3	0.27	0.27	0.05	0.31	0.14	0.88	0.5	0.6	2
0.26	0.31	0.3	0.3	0.04	0.31	0.14	0.88	0.5	0.6	1
0.28	0.32	0.28	0.28	0.05	0.35	0.14	0.88	0.5	0.6	1

Predicted Stock Price for DSE (Partial)

Date	Actual high price	Predicted high price
2016-02-14	575	567.104408
2016-02-15	570	565.41389
2016-02-16	570	557.9748441
2016-02-17	569.8	562.1146667
2016-02-18	567	560.5571734
2016-02-22	560.3	564.9972015
2016-02-23	562.7	560.4792054
2016-02-24	559	560.8838751
2016-02-25	560	565.5088182
2016-02-28	569.3	561.5189172
2016-02-29	566	564.5064042
2016-03-01	562	560.901061
2016-03-02	569.7	572.1518561
2016-03-03	570	554.4631048
2016-03-06	568.2	566.5079885
2016-03-07	566.8	563.109268
2016-03-08	564	555.9084887
2016-03-09	564.7	560.4712885

(continued)

(continued)

Date	Actual high price	Predicted high price
2016-03-10	563	562.3853059
2016-03-13	561.9	564.7234263
2016-03-14	560.2	561.5902217
2016-03-15	560.2	553.1495136
2016-03-16	559.8	563.2032919
2016-03-20	558.9	542.9006256
2016-03-21	558.2	548.4974759
2016-03-22	557.9	566.8374045
2016-03-23	560	561.1573945
2016-03-24	551.9	557.6789548

Predicted Stock Price for IBM, NYSE (Partial)

Date	High price	Predicted price
1-Nov-16	153.91	153.5217954
2-Nov-16	153.35	154.5623053
3-Nov-16	153.74	154.163205
4-Nov-16	153.64	156.0988626
7-Nov-16	156.11	156.6354739
8-Nov-16	155.93	155.7437583
9-Nov-16	155.56	160.7995985
10-Nov-16	161.16	161.1298414
11-Nov-16	161.34	161.9060894
14-Nov-16	161.86	158.6418793
15-Nov-16	159.15	159.62198
16-Nov-16	159.55	159.93521
17-Nov-16	159.93	160.5806973
18-Nov-16	160.72	162.819883
21-Nov-16	163	162.936417
22-Nov-16	163	161.9651804
23-Nov-16	162.38	162.8870071
25-Nov-16	163.19	164.9447125
28-Nov-16	164.66	164.2288648
29-Nov-16	164.41	163.3701992
30-Nov-16	163.8	162.4664055
1-Dec-16	162.2	160.9661825
2-Dec-16	160.29	161.8363297

(continued)

(continued)

Date	High price	Predicted price
5-Dec-16	161.15	161.47892
6-Dec-16	160.79	165.339742
7-Dec-16	165.18	166.5287985
8-Dec-16	166	167.2625337
9-Dec-16	166.72	167.8186317
12-Dec-16	166.79	171.961337
13-Dec-16	169.95	172.1560214
14-Dec-16	169.89	169.4907888
15-Dec-16	169.85	169.9466108
16-Dec-16	169.11	167.7368666
19-Dec-16	167.26	167.3167567
20-Dec-16	168.25	167.3502629
21-Dec-16	167.94	167.4708727

Predicted Stock Price for Microsoft, NASDAQ (Partial)

Date	High price	Predicted price
1-Nov-16	60.02	59.85341687
2-Nov-16	59.93	59.48643484
3-Nov-16	59.64	59.8107413
4-Nov-16	59.28	60.78194607
7-Nov-16	60.52	60.79773199
8-Nov-16	60.78	60.29013033
9-Nov-16	60.59	60.17522061
10-Nov-16	60.49	58.92571291
11-Nov-16	59.12	58.89426991
14-Nov-16	59.08	59.4716785
15-Nov-16	59.49	59.62661375
16-Nov-16	59.66	60.9383013
17-Nov-16	60.95	61.08024971
18-Nov-16	61.14	60.72042479
21-Nov-16	60.97	60.95745316
22-Nov-16	61.26	60.8435135
23-Nov-16	61.1	60.27370776
25-Nov-16	60.53	60.82941335
28-Nov-16	61.02	61.17455982
29-Nov-16	61.41	61.06014329

(continued)

(continued)

Date	High price	Predicted price
30-Nov-16	61.18	60.15983265
1-Dec-16	60.15	59.71492073
2-Dec-16	59.47	61.15964863
5-Dec-16	60.59	61.19285246
6-Dec-16	60.46	61.31860902
7-Dec-16	61.38	61.42687908
8-Dec-16	61.58	61.91230784
9-Dec-16	61.99	62.23494712
12-Dec-16	62.3	63.37615115
13-Dec-16	63.42	63.62880811
14-Dec-16	63.45	63.14978196
15-Dec-16	63.15	63.07456504
16-Dec-16	62.95	63.81526137
19-Dec-16	63.77	64.0740938
20-Dec-16	63.8	63.76901177
21-Dec-16	63.7	64.17915854
22-Dec-16	64.1	63.49788455

Predicted Stock Price for Digi, Bursa Malaysia (Partial)

Date	High price	Predicted price
20161117	5.05	5.026104957
20161118	5.02	4.99638383
20161121	4.99	4.997618784
20161122	4.99	5.027148684
20161123	5.02	5.030042152
20161124	5.01	5.008830135
20161125	5	5.012364769
20161128	5	5.018324306
20161129	5.01	5.002249116
20161130	5	4.995711232
20161201	4.99	4.985371102
20161202	4.98	4.996754893
20161205	4.98	4.961938902
20161206	4.97	4.9853446
20161207	4.99	4.979235465
20161208	4.96	4.986594763

(continued)

(continued)

Date	High price	Predicted price
20161209	4.98	4.982747736
20161213	4.99	4.988881997
20161214	4.99	5.000623984
20161215	4.99	5.023841968
20161216	5	5.009126594
20161219	5	4.996199664
20161220	5	5.020539845
20161221	4.99	5.008051237
20161222	4.99	4.971299992
20161223	4.97	5.022726837
20161227	5	5.013838595
20161228	5	5.037868758
20161229	5	5.037548222
20161230	5	5.022618618

Stock Investment Decision Data for IBM, NYSE (Partial)

Date	Open	Close	High	Low	Decision
12-Apr-16	149.61	149.63	150.11	148.58	Hold
11-Apr-16	150.26	149.25	151.95	149.22	Hold
8-Apr-16	149.06	149.35	149.99	148.12	Hold
7-Apr-16	149.07	148.25	149.6	148	Hold
6-Apr-16	149.9	150.02	150.15	147.81	Hold
5-Apr-16	151.43	150	151.6	149.65	Hold
4-Apr-16	152.34	152.07	153.52	151.91	Sell
1-Apr-16	150.51	152.52	152.96	150.25	Hold
31-Mar-16	149.79	151.45	153.1	149.36	Hold
30-Mar-16	149.95	148.41	150.41	148.32	Hold
29-Mar-16	147.59	149.33	149.76	147.5	Hold
28-Mar-16	147.75	148.4	148.65	147.23	Hold
24-Mar-16	144.98	147.95	148.22	144.49	Buy
23-Mar-16	148	145.4	148.03	145.13	Hold
22-Mar-16	148.06	148.1	149.28	147.84	Sell
21-Mar-16	147.3	148.63	148.71	146.72	Hold
18-Mar-16	147.4	147.09	147.51	145.51	Hold
17-Mar-16	144.78	147.04	147.32	144.45	Buy
16-Mar-16	142.62	144.79	144.88	142.11	Buy

(continued)

(continued)

Date	Open	Close	High	Low	Decision
15-Mar-16	141.74	142.96	143.33	141.54	Buy
14-Mar-16	142.01	142.78	143.19	141.04	Hold
11-Mar-16	141.73	142.36	142.92	140.51	Buy
10-Mar-16	141.24	140.19	141.47	138.09	Buy
9-Mar-16	139.31	140.41	142.17	139.23	Hold
8-Mar-16	139.71	139.07	140.35	137.42	Hold
7-Mar-16	137.28	140.15	140.51	136.87	Hold
4-Mar-16	137.54	137.8	139.42	137.02	Hold
3-Mar-16	137.22	137.8	137.96	136.07	Hold
2-Mar-16	133.7	136.3	137.44	133.22	Buy
1-Mar-16	132.24	134.37	134.64	132.03	Buy

Stock Investment Decision Data for IBM, NYSE (Partial)

Date	Open	High	Low	Close	Decision
13-Apr-16	55.12	55.35	55.44	54.89	Hold
12-Apr-16	54.37	54.65	54.78	53.76	Buy
11-Apr-16	54.49	54.31	55.15	54.3	Hold
8-Apr-16	54.67	54.42	55.28	54.32	Sell
7-Apr-16	54.87	54.46	54.91	54.23	Hold
6-Apr-16	54.36	55.12	55.2	54.21	Hold
5-Apr-16	55.19	54.56	55.3	54.46	Hold
4-Apr-16	55.43	55.43	55.66	55	Sell
1-Apr-16	55.05	55.57	55.61	54.57	Hold
31-Mar-16	54.95	55.23	55.59	54.86	Hold
30-Mar-16	54.93	55.05	55.64	54.9	Hold
29-Mar-16	53.66	54.71	54.86	53.45	Buy
28-Mar-16	54.21	53.54	54.29	53.33	Hold
24-Mar-16	53.84	54.21	54.33	53.73	Hold
23-Mar-16	54.11	53.97	54.24	53.74	Hold
22-Mar-16	53.61	54.07	54.25	53.46	Hold
21-Mar-16	53.25	53.86	53.93	52.93	Buy
18-Mar-16	54.92	53.49	54.97	53.45	Hold
17-Mar-16	54.21	54.66	55	54	Hold
16-Mar-16	53.45	54.35	54.6	53.4	Buy
15-Mar-16	52.75	53.59	53.59	52.74	Buy
14-Mar-16	52.71	53.17	53.59	52.63	Hold

(continued)

(continued)

Date	Open	High	Low	Close	Decision
11-Mar-16	53	53.07	53.07	52.38	Hold
10-Mar-16	52.93	52.05	52.94	51.16	Buy
9-Mar-16	51.89	52.84	52.85	51.86	Hold
8-Mar-16	50.8	51.65	52.13	50.6	Buy
7-Mar-16	51.56	51.03	51.8	50.58	Hold
4-Mar-16	52.4	52.03	52.45	51.71	Sell
3-Mar-16	52.97	52.35	52.97	51.78	Sell
2-Mar-16	52.41	52.95	52.96	52.16	Hold
1-Mar-16	50.97	52.58	52.59	50.92	Buy
29-Feb-16	51.35	50.88	51.65	50.66	Hold
26-Feb-16	52.6	51.3	52.68	51.1	Sell
25-Feb-16	51.73	52.1	52.1	50.61	Hold
24-Feb-16	50.69	51.36	51.5	50.2	Buy
23-Feb-16	52.34	51.18	52.37	50.98	Hold
22-Feb-16	52.28	52.65	53	52.28	Hold
19-Feb-16	51.97	51.82	52.28	51.53	Buy

Stock Investment Decision Data for Digi, Bursa Malaysia (Partial)

Date	Open	High	Low	Close	Decision
20161111	4.97	5.02	4.91	4.99	Buy
20161114	4.99	4.99	4.88	4.92	Hold
20161115	4.92	4.99	4.92	4.98	Hold
20161116	5	5.02	4.98	4.99	Sell
20161117	5	5.01	4.93	5	Hold
20161118	4.99	5	4.95	4.99	Hold
20161121	5	5	4.97	4.99	Hold
20161122	5	5.01	4.94	4.99	Hold
20161123	5	5	4.96	4.99	Hold
20161124	4.99	4.99	4.95	4.97	Buy
20161125	4.95	4.98	4.95	4.97	Hold
20161128	4.97	4.98	4.96	4.97	Hold
20161129	4.93	4.97	4.9	4.95	Buy
20161130	4.97	4.99	4.87	4.87	Hold
20161201	4.9	4.96	4.88	4.95	Hold
20161202	4.95	4.98	4.94	4.95	Hold
20161205	4.91	4.99	4.91	4.96	Hold

(continued)

(continued)

Date	Open	High	Low	Close	Decision
20161206	4.97	4.99	4.95	4.98	Hold
20161207	4.97	4.99	4.96	4.98	Hold
20161208	4.99	5	4.97	4.99	Sell
20161209	4.95	5	4.95	4.99	Hold
20161213	5	5	4.96	4.98	Hold
20161214	4.96	4.99	4.96	4.98	Hold
20161215	4.94	4.99	4.93	4.97	Hold
20161216	4.94	4.97	4.94	4.96	Hold
20161219	4.94	5	4.94	4.98	Hold
20161220	5	5	4.97	4.99	Hold
20161221	5	5	4.97	4.98	Hold
20161222	4.99	5	4.95	4.97	Hold
20161223	4.97	4.99	4.95	4.95	Hold
20161227	4.93	4.97	4.93	4.93	Buy
20161228	4.93	4.99	4.93	4.94	Hold
20161229	4.94	4.98	4.92	4.96	Hold
20161230	4.95	4.97	4.83	4.83	Hold

Stock Price Prediction Validation Data for ACI, DSE (Partial)

Date	High price	Predicted high price	Forecasting error (%)	MAPE	MAD	RMSE
2016-02-14	575	567.104408	1.373146443	3.337062276	7.895592045	62.34037374
2016-02-15	570	565.41389	0.804580696		4.586109968	21.03240464
2016-02-16	570	557.9748441	2.109676467		12.02515586	144.6043735
2016-02-17	569.8	562.1146667	1.348777343		7.685333298	59.0643479
2016-02-18	567	560.5571734	1.13630099		6.442826612	41.51001475
2016-02-22	560.3	564.9972015	0.838336878		4.697201526	22.06370217
2016-02-23	562.7	560.4792054	0.394667599		2.220794579	4.931928563
2016-02-24	559	560.8838751	0.337008075		1.883875138	3.548985536
2016-02-25	560	565.5088182	0.983717529		5.508818162	30.34707755
2016-02-28	569.3	561.5189172	1.36678074		7.781082754	60.54524883
2016-02-29	566	564.5064042	0.263886176		1.493595757	2.230828286
2016-03-01	562	560.901061	0.195540742		1.098938971	1.207666862
2016-03-02	569.7	572.1518561	0.430376713		2.451856133	6.011598498
2016-03-03	570	554.4631048	2.725771083		15.53689517	241.3951116
2016-03-06	568.2	566.5079885	0.297784488		1.692011462	2.862902787
2016-03-07	566.8	563.109268	0.651152437		3.690732013	13.62150279

(continued)

(continued)

Date	High price	Predicted high price	Forecasting error (%)	MAPE	MAD	RMSE
2016-03-08	564	555.9084887	1.434665126		8.091511309	65.47255526
2016-03-09	564.7	560.4712885	0.748842122		4.228711463	17.88200064
2016-03-10	563	562.3853059	0.109181907		0.614694136	0.37784888
2016-03-13	561.9	564.7234263	0.502478433		2.823426317	7.971736168
2016-03-14	560.2	561.5902217	0.24816525		1.39022173	1.932716459
2016-03-15	560.2	553.1495136	1.258565939		7.050486388	49.7093583
2016-03-16	559.8	563.2032919	0.607947817		3.403291879	11.58239562
2016-03-20	558.9	542.9006256	2.862654209		15.99937438	255.9799804
2016-03-21	558.2	548.4974759	1.738180598		9.702524096	94.13897384
2016-03-22	557.9	566.8374045	1.601972489		8.937404516	79.87719948
2016-03-23	560	561.1573945	0.206677593		1.157394519	1.339562074
2016-03-24	551.9	557.6789548	1.047101787		5.778954763	33.39631815
2016-03-27	555	552.4391825	0.461408566		2.560817539	6.557786466
2016-03-28	542.6	536.8608295	1.057716636		5.739170466	32.93807764
2016-03-29	535.2	548.5876587	2.501430995		13.38765869	179.2294051
2016-03-30	542	551.7744488	1.80340384		9.774448813	95.5398496
2016-03-31	549	547.443365	0.283540078		1.55663503	2.423112615
2016-04-03	546.5	550.4843579	0.729068237		3.984357917	15.87510801
2016-04-04	558.7	568.4842289	1.75124913		9.784228891	95.73113499
2016-04-05	560.8	574.8960249	2.513556505		14.09602488	198.6979174
2016-04-06	556	573.449396	3.138380572		17.44939598	304.48142
2016-04-07	574	581.9066801	1.377470396		7.906680075	62.51558982

Stock Price Prediction Validation Data for Bursa Malaysia (Partial)

Date	High price	Predicted high price	Forecasting error (%)	MAPE	MAD	RMSE
20161117	5.05	5.026104957	0.473169174	0.796920921	0.023895	0.00057097
20161118	5.02	4.99638383	0.470441628		0.023616	0.00055772
20161121	4.99	4.997618784	0.152681041		0.007619	5.8046E−05
20161122	4.99	5.027148684	0.744462599		0.037149	0.00138002
20161123	5.02	5.030042152	0.200042871		0.010042	0.00010084
20161124	5.01	5.008830135	0.023350599		0.00117	1.3686E−06
20161125	5	5.012364769	0.247295374		0.012365	0.00015289
20161128	5	5.018324306	0.366486116		0.018324	0.00033578
20161129	5.01	5.002249116	0.154708272		0.007751	6.0076E−05
20161130	5	4.995711232	0.085775369		0.004289	1.8394E−05

(continued)

(continued)

Date	High price	Predicted high price	Forecasting error (%)	MAPE	MAD	RMSE
20161201	4.99	4.985371102	0.092763485		0.004629	2.1427E−05
20161202	4.98	4.996754893	0.336443628		0.016755	0.00028073
20161205	4.98	4.961938902	0.362672655		0.018061	0.0003262
20161206	4.97	4.9853446	0.308744464		0.015345	0.00023546
20161207	4.99	4.979235465	0.215722151		0.010765	0.00011588
20161208	4.96	4.986594763	0.536184732		0.026595	0.00070728
20161209	4.98	4.982747736	0.055175424		0.002748	7.5501E−06
20161213	4.99	4.988881997	0.022404862		0.001118	1.2499E−06
20161214	4.99	5.000623984	0.212905486		0.010624	0.00011287
20161215	4.99	5.023841968	0.678195753		0.033842	0.00114528
20161216	5	5.009126594	0.182531888		0.009127	8.3295E−05
20161219	5	4.996199664	0.076006718		0.0038	1.4443E−05
20161220	5	5.020539845	0.410796892		0.02054	0.00042189
20161221	4.99	5.008051237	0.361748227		0.018051	0.00032585
20161222	4.99	4.971299992	0.374749661		0.0187	0.00034969
20161223	4.97	5.022726837	1.060902145		0.052727	0.00278012
20161227	5	5.013838595	0.27677191		0.013839	0.00019151
20161228	5	5.037868758	0.757375157		0.037869	0.00143404
20161229	5	5.037548222	0.750964442		0.037548	0.00140987
20161230	5	5.022618618	0.452372362		0.022619	0.0005116

Stock Price Prediction Validation Data for IBM, NYSE (Partial)

Date	High price	Predicted high price	Forecasting error (%)	MAPE	MAD	RMSE
1-Mar-16	134.64	137.2290225	1.922922254	0.87053416	2.589023	6.703038
2-Mar-16	137.44	137.542491	0.074571433		0.102491	0.010504
3-Mar-16	137.96	139.3868376	1.034240041		1.426838	2.035865
4-Mar-16	139.42	140.5829261	0.834117112		1.162926	1.352397
7-Mar-16	140.51	140.4626231	0.03371778		0.047377	0.002245
8-Mar-16	140.35	142.5220042	1.547562676		2.172004	4.717602
9-Mar-16	142.17	141.6724808	0.349946655		0.497519	0.247525
10-Mar-16	141.47	143.3131481	1.302854363		1.843148	3.397195
11-Mar-16	142.92	143.8282358	0.63548546		0.908236	0.824892
14-Mar-16	143.19	144.5345353	0.938986903		1.344535	1.807775
15-Mar-16	143.33	146.4850069	2.201218807		3.155007	9.954069

(continued)

(continued)

Date	High price	Predicted high price	Forecasting error (%)	MAPE	MAD	RMSE
16-Mar-16	144.88	147.6506893	1.912402915		2.770689	7.676719
17-Mar-16	147.32	148.0548367	0.498803099		0.734837	0.539985
18-Mar-16	147.51	149.63284	1.439116007		2.12284	4.50645
21-Mar-16	148.71	150.5790477	1.25684061		1.869048	3.493339
22-Mar-16	149.28	148.1899917	0.730177018		1.090008	1.188118
23-Mar-16	148.03	148.4939516	0.313417289		0.463952	0.215251
24-Mar-16	148.22	148.7364049	0.348404361		0.516405	0.266674
28-Mar-16	148.65	150.0443661	0.938019574		1.394366	1.944257
29-Mar-16	149.76	150.9063695	0.765471063		1.146369	1.314163
30-Mar-16	150.41	153.5053872	2.057966377		3.095387	9.581422
31-Mar-16	153.1	153.6858524	0.382659959		0.585852	0.343223
1-Apr-16	152.96	153.960154	0.653866371		1.000154	1.000308
4-Apr-16	153.52	152.3130722	0.786169768		1.206928	1.456675
5-Apr-16	151.6	150.8597119	0.488316721		0.740288	0.548027
6-Apr-16	150.15	149.7800794	0.246367367		0.369921	0.136841
7-Apr-16	149.6	150.3084755	0.473579847		0.708475	0.501937
8-Apr-16	149.99	152.7064381	1.811079457		2.716438	7.379036
11-Apr-16	151.95	150.60866	0.882750927		1.34134	1.799193
12-Apr-16	150.11	151.7115356	1.066908003		1.601536	2.564916

Stock Price Prediction Validation Data for Microsoft, NASDAQ (Partial)

Date	High price	Predicted high price	Forecasting error (%)	MAPE	MAD	RMSE
19-Feb-16	52.28	53.10262496	1.573498388	1.111934817	0.822625	0.676712
22-Feb-16	53	52.2440023	1.426410761		0.755998	0.571533
23-Feb-16	52.37	51.6051487	1.460476032		0.764851	0.584998
24-Feb-16	51.5	52.20588127	1.370643235		0.705881	0.498268
25-Feb-16	52.1	52.69330547	1.138782083		0.593305	0.352011
26-Feb-16	52.68	51.6567134	1.942457486		1.023287	1.047115
29-Feb-16	51.65	52.65354016	1.942962557		1.00354	1.007093
1-Mar-16	52.59	53.02727629	0.831481826		0.437276	0.191211
2-Mar-16	52.96	53.00669357	0.088167614		0.046694	0.00218
3-Mar-16	52.97	52.48310571	0.919188768		0.486894	0.237066
4-Mar-16	52.45	51.61296855	1.595865482		0.837031	0.700622
7-Mar-16	51.8	52.21337489	0.798021021		0.413375	0.170879
8-Mar-16	52.13	52.88401572	1.446414199		0.754016	0.56854

(continued)

(continued)

Date	High price	Predicted high price	Forecasting error (%)	MAPE	MAD	RMSE
9-Mar-16	52.85	52.99067799	0.266183523		0.140678	0.01979
10-Mar-16	52.94	53.04336139	0.195242516		0.103361	0.010684
11-Mar-16	53.07	53.78959417	1.355934001		0.719594	0.517816
14-Mar-16	53.59	53.67298201	0.154846078		0.082982	0.006886
15-Mar-16	53.59	54.36045455	1.437683427		0.770455	0.5936
16-Mar-16	54.6	55.08867958	0.895017538		0.48868	0.238808
17-Mar-16	55	54.87876033	0.220435772		0.12124	0.014699
18-Mar-16	54.97	53.86385751	2.012265754		1.106142	1.223551
21-Mar-16	53.93	54.23100819	0.558146097		0.301008	0.090606
22-Mar-16	54.25	54.33072316	0.148798444		0.080723	0.006516
23-Mar-16	54.24	54.20015017	0.073469459		0.03985	0.001588
24-Mar-16	54.33	54.32746599	0.004664105		0.002534	6.42E−06
28-Mar-16	54.29	54.97089401	1.254179428		0.680894	0.463617
29-Mar-16	54.86	55.45412409	1.082982297		0.594124	0.352983
30-Mar-16	55.64	55.49076035	0.268223678		0.14924	0.022272
31-Mar-16	55.59	55.52272915	0.121012506		0.067271	0.004525
1-Apr-16	55.61	55.69015391	0.144135787		0.080154	0.006425

Stock Price with Factors Data to Measure the Affect for DSE (Partial)

Low	High	Average	Close	Gold price	Dollar price	Bank interest rate	FDI	Inflation
499.00	510.90	507.03	507.07	2905.00	69.25	8.50	861,736,237.16	9.00
490.00	510.00	504.89	504.76	2905.00	69.25	8.50	861,736,237.16	9.00
478.00	504.00	500.83	500.71	2905.00	69.24	8.50	861,736,237.16	9.00
490.00	502.80	499.15	499.16	2905.00	69.24	8.50	861,736,237.16	9.00
490.00	508.00	502.03	502.23	2905.00	69.24	8.50	861,736,237.16	9.00
501.00	505.00	502.69	502.71	2905.00	69.27	8.50	861,736,237.16	9.00
496.10	505.00	499.16	499.17	2905.00	69.26	8.50	861,736,237.16	9.00
490.50	499.80	494.08	493.96	2905.00	69.25	8.50	861,736,237.16	9.00
480.00	500.00	491.54	491.51	2905.00	69.26	8.50	861,736,237.16	9.00
470.00	495.00	491.04	491.13	2905.00	69.26	8.50	861,736,237.16	9.00
480.00	500.00	497.24	497.19	2905.00	69.25	8.50	861,736,237.16	9.00
491.10	495.50	492.41	492.43	2905.00	69.25	8.50	861,736,237.16	9.00
490.00	495.00	491.55	491.57	2905.00	69.25	8.50	861,736,237.16	9.00
443.20	498.00	494.66	495.08	2905.00	69.25	8.50	861,736,237.16	9.00

(continued)

(continued)

Low	High	Average	Close	Gold price	Dollar price	Bank interest rate	FDI	Inflation
491.00	501.90	497.04	497.14	2905.00	69.25	8.50	861,736,237.16	9.00
445.60	495.00	489.34	489.68	2905.00	69.25	8.50	861,736,237.16	9.00
480.00	489.00	486.30	486.24	2905.00	69.25	8.50	861,736,237.16	9.00
460.00	489.00	479.42	479.44	2905.00	69.25	8.50	861,736,237.16	9.00
451.00	475.10	457.10	457.13	2905.00	69.25	8.50	861,736,237.16	9.00
440.00	459.00	454.73	454.65	2905.00	69.25	8.50	861,736,237.16	9.00
430.00	452.00	443.94	444.02	2905.00	69.25	8.50	861,736,237.16	9.00
425.00	445.00	435.57	435.57	2905.00	69.25	8.50	861,736,237.16	9.00
400.00	456.80	440.84	441.22	2905.00	69.25	8.50	861,736,237.16	9.00
438.00	445.00	440.67	440.57	2995.00	69.25	8.50	861,736,237.16	9.00
438.40	442.00	440.11	440.09	2995.00	69.26	8.50	861,736,237.16	9.00
421.50	433.00	426.42	426.33	2995.00	69.31	8.50	861,736,237.16	9.00
400.00	440.00	437.12	437.19	2995.00	69.30	8.50	861,736,237.16	9.00
425.00	440.00	435.05	434.56	2995.00	69.30	8.50	861,736,237.16	9.00
419.90	435.00	430.96	430.99	2995.00	69.31	8.50	861,736,237.16	9.00
410.00	431.90	429.24	429.31	2995.00	69.31	8.50	861,736,237.16	9.00
400.00	428.50	418.99	419.15	2995.00	69.29	8.50	861,736,237.16	9.00

Bibliography

Goyal S, Patterh MS (2016) Modified bat algorithm for localization of wireless sensor network. Wireless Pers Commun 86(2):657–670

Kavousi-Fard A, Samet H, Marzbani F (2014) A new hybrid modified firefly algorithm and support vector regression model for accurate short term load forecasting. Expert Syst Appl 41(13):6047–6056

Rather AM, Agarwal A, Sastry V (2015) Recurrent neural network and a hybrid model for prediction of stock returns. Expert Syst Appl 42(6):3234–3241

Song X, Tang L, Zhao S, Zhang X, Li L, Huang J, Cai W (2015) Grey wolf optimizer for parameter estimation in surface waves. Soil Dyn Earthq Eng 75:147–157

D. Das et al., *Optimization Algorithms in Machine Learning*, Engineering Optimization: Methods and Applications, https://doi.org/10.1007/978-981-96-3849-9

The manufacturer's authorised representative in the EU is Springer Nature Customer Service Centre GmbH, Europaplatz 3, 69115 Heidelberg, Germany. If you have any concerns regarding our products, please contact ProductSafety@springernature.com

Printed and bound by CPI Group (UK) Ltd, Croydon, CR0 4YY
30/09/2026
02225265-0015